T0297414

Complexity-Aware High Efficiency Video Coding

Guilherme Corrêa • Pedro Assunção
Luciano Agostini • Luis A. da Silva Cruz

Complexity-Aware High Efficiency Video Coding

Guilherme Corrêa
Computação
Centro de Desenvolvimento Tecnológico,
Universidade Federal de Pelotas
Pelotas, Brazil

Luciano Agostini
Computação
Centro de Desenvolvimento Tecnológico,
Universidade Federal de Pelotas
Pelotas, Brazil

Pedro Assunção
Instituto de Telecomunicações (IT)
Instituto Politécnico de Leiria
Leiria, Portugal

Luis A. da Silva Cruz
Instituto de Telecomunicações (IT)
Dep. Eng. Electrotécnica e de
Computadores
Faculdade de Ciências e Tecnologia
Universidade de Coimbra
Coimbra, Portugal

ISBN 978-3-319-25776-1 ISBN 978-3-319-25778-5 (eBook)
DOI 10.1007/978-3-319-25778-5

Library of Congress Control Number: 2015954576

Springer Cham Heidelberg New York Dordrecht London

Printed on acid-free paper

Springer International Publishing AG Switzerland is part of Springer Science+Business Media (www.springer.com)

Preface

In the last decades, the rapid advances of semiconductor technologies fostered a large development in the field of multimedia systems, mainly due to the continuous increase of computational resources and the availability of reliable communication infrastructures. Several video compression standards have been developed in this period, aiming at reducing transmission bit rates without decreasing the video quality. The High-Efficiency Video Coding (HEVC) standard, recently launched by the Joint Collaborative Team on Video Coding (JCT-VC), is the state of the art in video compression and is expected to gradually substitute its predecessor, the H.264/AVC standard. HEVC provides improved compression ratios in comparison to previous standards, but such gains are associated with large increases in the encoding computational complexity and consequently longer processing times, which may compromise the encoder operation in portable devices and in real-time systems, especially for high-resolution videos. This book addresses the subject of computational complexity of HEVC encoders with contributions extending from the analysis of HEVC compression efficiency and computational complexity to the reduction and scaling of its encoding complexity. Besides the introductory chapters, which present an overview of the HEVC standard and the state-of-the-art works on the field, this book also introduces four main contributions from the authors, which target on analysing and solving complexity-related issues in HEVC encoders. The first contribution is an investigation and detailed analysis of the HEVC encoding tools which allowed identifying the most computationally demanding operations of the encoding process. The second contribution of this book comprises a set of five new algorithms for reducing and dynamically scaling the encoding complexity of HEVC encoders. All of them take advantage from the flexibility of the frame partitioning structures allowed by the standard, namely, the coding units and the prediction units, which were identified as responsible for a large share of the encoding computational complexity. The best complexity scaling algorithm presented in this book allows downscaling the encoding complexity to 50 % of its original value with negligible loss of compression efficiency and down to 20 % with medium to small loss. The third book contribution consists of a set of early termination methods based on data mining techniques, which are able to reduce the computational complexity required to find the best frame partitioning structures, namely, the coding trees, the prediction units and the residual quadtrees, in up to 65 % with very small compression efficiency loss. Finally, the fourth contribution of this book is an encoding time control system that employs the three previous contributions to adjust the encoding time whenever necessary and maintain it under a specified target. The system uses predefined encoding configurations created by combining the early termination schemes and by changing the parameterisation of the most computationally demanding tools of HEVC. Overall, the methods proposed in this book are especially useful in power-constrained portable multimedia devices to reduce energy consumption and to extend the battery life. Besides, they can also be applied to portable and non-portable multimedia devices operating in real time with limited computational resources.

Pelotas, Brazil — Guilherme Corrêa
Leiria, Portugal — Pedro Assunção
Pelotas, Brazil — Luciano Agostini
Coimbra, Portugal — Luis A. da Silva Cruz

Contents

Abbreviations

AI	All intra
ALF	Adaptive loop filter
AMP	Asymmetric motion partition
ARFF	Attribute-relation file format
AVC	Advanced video coding
BD	Bjøntegaard delta
BD-rate	Bjøntegaard delta bit rate
BD-PSNR	Bjøntegaard delta PSNR
CABAC	Context-adaptive binary arithmetic coding
CB	Coding block
CBF	Coded block flag
CCUPU	Constrained coding units and prediction units
CIF	Common intermediate format
CTB	Coding tree block
CTC	Common test conditions
CTDE	Coding tree depth estimation
CTU	Coding tree unit
CU	Coding unit
DBF	Deblocking filter
DCT	Discrete cosine transform
DMV	Differential motion vector
DPB	Decoded picture buffer
DST	Discrete sine transform
EPZS	Enhanced predictive zonal search
Fc	Constrained frame
FDCR	Fixed depth complexity reduction
FME	Fast motion estimation
fps	Frames per second
FRME	Fractional motion estimation
FS	Full search
Fu	Unconstrained frame

GBFOS Generalised Breiman, Friedman, Olshen and Stone Algorithm
GOP Group of pictures
GPB Generalised P and B picture
HD High definition
HE High efficiency
HETR High-efficiency encoding time reduction
HEVC High-efficiency video coding
HM HEVC model
HP High profile
I/O Input/output
IBD Internal bit depth
IBDD Internal bit depth decrease
IBDI Internal bit depth increase
IDR Instantaneous decoding refresh
IGAE Information gain attribute evaluation
IEC International electrotechnical commission
IME Integer motion estimation
IP Intra-prediction
IQ Inverse quantisation
ISO International organisation for standardisation
IT Inverse transform
JCT-VC Joint collaborative team on video coding
KDD Knowledge discovery from data
KLD Kullback-Leibler divergence
LC Low complexity
LCB Largest coding block
LCTC Low-complexity encoding time control
LD Low delay
LDP Low delay P
LM Linear mode
MB Macroblock
MC Motion compensation
MCTDL Motion-compensated tree depth limitation
MD Mode decision
ME Motion estimation
MPEG Moving Picture Experts Group
MPM Most probable modes
MSE Mean-squared error
MSM Merge/SKIP mode
MTDM Maximum tree depth map
MV Motion vector
MVP Motion vector prediction
NSQT Non-square transforms
PB Prediction block
PRD Power-rate-distortion

PRDO	Power-rate-distortion optimisation
PSNR	Peak signal-to-noise ratio
PU	Prediction unit
Q	Quantisation
QP	Quantisation parameter
QZB	Quasi-zero-blocks
RA	Random access
RAP	Random access point
R-D	Rate-distortion
R-D-C	Rate-distortion-complexity
RDCO	Rate-distortion-complexity optimisation
RDO	Rate-distortion optimisation
RDOQ	Rate-distortion optimised quantisation
RGB	Red, green and blue
RMD	Rough mode decision
RQT	Residual quadtree
SA	Search area
SAD	Sum of absolute differences
SAO	Sample adaptive offset
SATD	Sum of absolute transformed differences
SCB	Smallest coding block
SMP	Symmetric motion partition
SSE	Sum of squared error
SUMHexS	Simple unsymmetrical-cross multi-hexagon-grid search
T	Transform
TB	Transform block
TSS	Three-step search
TU	Transform unit
TZS	Test zone search
UMHexS	Unsymmetrical-cross multi-hexagon-grid search
VCEG	Video coding experts group
VDCR	Variable depth complexity reduction
VGA	Video graphics array
YCbCr	Luminance, blue chrominance and red chrominance
Y-PSNR	Luminance PSNR

List of Figures

List of Tables

Chapter 1
Introduction

In the last decades, the fast advances of semiconductor technologies fostered a large expansion in the consumer market of multimedia-ready devices due to the continuous increase of computational resources and availability of reliable communication infrastructures. Digital video has been available since the beginning of the so-called information age, but only after the expansion and popularisation of personal computers and high-speed networks did this type of information become so widely present in our daily lives.

Nowadays, digital televisions, portable computers, personal digital assistants (PDA) and mobile phones are among the most popular consumer equipment able to receive and display high-resolution video in real time. Very common are also those devices that can capture and transmit digital video through wired and wireless channels. Furthermore, the current trend in most portable devices with embedded digital cameras is to include the capability of encoding and decoding high-resolution digital video streams.

Despite the recent evolution in portable devices, particularly in terms of communications technology and computational power, the limited battery capacity still imposes major constraints in multimedia applications demanding high computational power, such as those dealing with video encoding. In such cases, the user experience might be limited by the reduced battery capacity. Furthermore, even in those cases in which battery is not an issue, encoding and decoding high-resolution digital video streams in real time is still a challenge, especially when considering the computational requirements of the most recent video coding standards.

Previous studies which examine typical use scenarios of portable devices have shown that a very significant amount of power consumption (from 40 to 60 %) is related to video encoding and decoding operations, with the encoder typically requiring the largest share of computing time and power consumption [1–4]. It has been claimed in [5] that more than two-thirds of this computational complexity corresponds

G. Corrêa et al., *Complexity-Aware High Efficiency Video Coding*,
DOI 10.1007/978-3-319-25778-5_1

to the encoding process, whereas the rest corresponds to transmission and input/output (I/O) operations. Even though the results presented in [5] are just an estimate for low-resolution video, the recent adoption of higher resolutions increased even more the computational needs of the video encoding processes, since greater computational efforts are necessary to process the increasingly larger amounts of video information. Furthermore, a consequence of current video coding standard evolution is that the use of more efficient signal processing tools is significantly augmenting the number of operations-per-pixel required in the newest video codecs.

This is the case of the state-of-the-art standard, the high-efficiency video coding (HEVC) [6], finished on March of 2013 by the Joint Collaborative Team on Video Coding (JCT-VC), from the International Telecommunication Union (ITU) [7] and the Joint Technical Committee 1 of the International Organization for Standardization and the International Electrotechnical Commission (ISO/IEC JTC1) [8]. HEVC achieves 40–50 % bit rate reduction in comparison with its predecessor, the H.264/AVC video coding standard [9], at the same subjective image quality. However, to reach such goal, HEVC incorporates several new tools, which increased the encoding computational complexity in a range from 9 to 502 % in comparison to H.264/AVC high profile [10].

Even though several works addressing complexity-aware video coding have been proposed in the last years and designed for use in previous video encoding standards, most of them cannot be directly applied to the particular case of HEVC, which is based on encoding structures and tools quite different from those of previous standards, as shown later in this book. This fact has kept the research on the field continuously active, since the demand for new strategies to reduce, scale and control the computational complexity of video encoders is renewed with the introduction of a much more complex video coding standard.

This book presents an overview of the HEVC standard and of the state-of-the-art works found in the literature that target at solving complexity-related issues in video encoders. Besides, the book presents the main authors' contribution on the field, which span from a performance and complexity assessment of the HEVC encoder to methods for reducing, scaling and controlling the encoding computational complexity while still maintaining a compromise with compression efficiency.

1.1 Terminology for Computational Complexity

As complexity does not have a single universal meaning and is sometimes a confounding concept, its definition in the context of this book is provided in this section, as well as the related concepts of complexity reduction and scaling.

Computational complexity is a term used to describe the amount of calculations performed in a task. In the specific case of the research field addressed in this book, computational complexity refers to the calculations performed in the whole video coding process or in a specific part of it. As the number of computations affects directly the total time of processor usage, computational complexity is always measured in terms of processing time in the experiments presented in this book, unless explicitly stated otherwise. In order to simplify the text, the term complexity is sometimes used as a synonym for computational complexity throughout the book.

Computational complexity reduction is a term used to describe those methods that yield fixed decreases in the computational complexity of a task. Once applied to a determined algorithm, a computational complexity reduction technique is able to decrease the amount of computational resources required to complete the encoding process to a level that is dependent upon the video source characteristics. In this context, the amount of complexity reduction achieved is unknown until the completeness of the task.

Computational complexity scaling is the process of adaptively adjusting the computational complexity of a task in order to reach a desired target complexity, which can be defined by a user or a system (e.g. operating system or transmission equipment). Complexity scaling methods are able to decrease or increase the computational effort employed in video encoding by adjusting the encoder parameters on the fly until the computational complexity is below a given upper limit. A feedback mechanism is generally used in such cases to guide the adjustment steps that are applied to reach the target complexity.

1.2 Book Organisation and Contribution

Chapter 2 presents an overview of video coding and decoding technology, including the basic concepts of digital video compression, the general operation of a video compression system, a detailed description of HEVC and discussions on RDO and computational complexity of video encoding.

Chapter 3 presents an overview of the state-of-the-art research on computational complexity reduction and scaling techniques for video encoding systems. A description of current methods for modelling, reducing and scaling the expenditure of computational resources on video codecs is presented along with future trends on complexity management for video codecs implemented in power-constrained or computationally constrained devices.

Chapters 4, 5, 6 and 7 present the main authors' contribution in the field, all of which focus on computational complexity analysis, scaling, reduction and encoding time control, respectively, as briefly described in the following subsections.

Finally, Chap. 8 presents the conclusions of this book, summarising the major results available in the literature, identifying some possible extensions and pointing out future research directions in the area.

1.2.1 Encoding Performance and Complexity Assessment

When developing complexity-aware video coding systems, one should focus on those tools or processes that are the most computationally intensive and look for alternatives or adaptations that yield good encoding performance at the cost of a smaller computational complexity. Therefore, a computational complexity assessment on the encoder and its tools and processes is essential prior to the development of such systems.

Chapter 4 presents an experimental investigation that was carried out in order to identify the tools that most affect the encoding efficiency and computational complexity of the HEVC encoder [10]. A set of encoding configurations was created to investigate the impact of each tool, varying the encoding parameters and comparing the results with a baseline encoder. The results of this study provided relevant information to implement complexity-constrained encoders by taking into account the trade-off between complexity and coding efficiency. Chapter 4 also presents a second research study that was carried out in order to identify the influence of the frame partitioning structures of HEVC in both computational complexity and coding efficiency [11]. The results of this study showed that the nature of the partitioning structures used in HEVC leads to nested rate-distortion optimisation (RDO) loops, which is the main contributing factor to the increased encoding computational complexity. Part of this research served as support to the development of complexity reduction and scaling solutions presented later in the book.

1.2.2 Algorithms for Computational Complexity Scaling

Chapter 5 presents a set of algorithms for computational complexity scaling based on dynamic adjustment of frame partitioning structures, namely, the coding tree units (CTUs), coding units (CUs) and the prediction units (PUs). All algorithms aim at adjusting the computational effort employed to decide the best frame partitioning structures according to a target computational complexity, which is limited by the amount of computational resources available in the encoder. Five approaches have been proposed:

- Fixed depth complexity scaling (FDCS) [11]
- Variable depth complexity scaling (VDCS) [12, 13]
- Motion-compensated tree depth limitation (MCTDL) [14]
- Coding tree depth estimation (CTDE) [15, 16]
- Constrained coding units and prediction units (CCUPU) [17, 18]

1.2.3 Data Mining for Computational Complexity Reduction

Chapter 6 introduces a set of procedures for deciding whether the partition structure decision should be terminated early or run to the end of an exhaustive search for the best configuration. The proposed methods are all based on decision trees obtained through data mining (DM) techniques. By extracting intermediate data from the encoding process, three sets of decision trees were devised to avoid running the RDO algorithm to its full extent:

- Early termination for determining coding trees [19, 20]
- Early termination for determining prediction units [20–22]
- Early termination for determining residual quadtrees

1.2.4 Complexity Reduction and Scaling Applied to Encoding Time Control

Chapter 7 presents an encoding time control algorithm for HEVC, which makes use of the contributions presented in previous chapters of the book to dynamically adjust the encoding process, maintaining the encoding time per group of pictures (GOP) under a determined upper bound. The CCUPU method mentioned in Sect. 1.2.2, the techniques for complexity reduction mentioned in Sect. 1.2.3 and the encoder configurations that most affect the computational complexity of HEVC, identified in the analysis mentioned in Sect. 1.2.1, were combined and applied to the development of a rate-distortion-complexity (R-D-C) optimised system that provides encoding time control for HEVC encoders. By adjusting the encoder operating point according to the best performing configurations identified in an extensive R-D-C efficiency analysis, the system provides encoding time control of medium to fine granularity.

References

1. W. Kim, J. You, J. Jeong, Complexity control strategy for real-time H.264/AVC encoder. IEEE Trans. Consum. Electron. **56**, 1137–1143 (2010)
2. S.B. Solak, F. Labeau, Complexity scalable video encoding for power-aware applications, in *2010 International Green Computing Conference* (2010), pp. 443–449.
3. L. Xiaoan, W. Yao, E. Erkip, Power efficient H.263 video transmission over wireless channels. IEEE Int. Conf. Image Process. **1**, I-533–I-536 (2002)
4. P. Agrawal, C. Jyh-Cheng, S. Kishore, P. Ramanathan, K. Sivalingam, Battery power sensitive video processing in wireless networks, in *9th IEEE International Symposium on Personal, Indoor and Mobile Radio Communications*, vol. 1 (1998), pp. 116–120.
5. A.K. Katsaggelos, Z. Fan, Y. Eisenberg, R. Berry, Energy-efficient wireless video coding and delivery. IEEE Wirel. Commun. **12**, 24–30 (2005)
6. G.J. Sullivan, J. Ohm, H. Woo-Jin, T. Wiegand, Overview of the High Efficiency Video Coding (HEVC) Standard. IEEE Trans. Circ. Syst. Video Technol. **22**, 1649–1668 (2012)
7. International Telecommunication Union, ITU-T Recommendation H.264 (05/2003): advanced video coding for generic audiovisual services (2003).
8. C.S. Kannangara, I.E. Richardson, A.J. Miller, Computational complexity management of a real-time H.264/AVC encoder. IEEE Trans. Circ. Syst. Video Technol. **18**, 1191–1200 (2008)
9. T. Wiegand, G.J. Sullivan, G. Bjontegaard, A. Luthra, Overview of the H.264/AVC video coding standard. IEEE Trans. Circ. Syst. Video Technol. **13**, 560–576 (2003)
10. G. Correa, P. Assuncao, L. Agostini, L.A. da Silva Cruz, Performance and computational complexity assessment of high efficiency video encoders. IEEE Trans. Circ. Syst. Video Technol. **22**, 1899–1909 (2012)
11. G. Correa, P. Assuncao, L. Agostini, L.A. da Silva Cruz, Complexity control of high efficiency video encoders for power-constrained devices. IEEE Trans. Consum. Electron. **57**, 1866–1874 (2011)

12. G. Correa, P. Assuncao, L.A. Da Silva Cruz, L. Agostini, Adaptive coding tree for complexity control of high efficiency video encoders, in *2012 Picture Coding Symposium* (2012), pp. 425–428.
13. G. Correa, P. Assuncao, L. A. da Silva Cruz, L. Agostini, Dynamic tree-depth adjustment for low power HEVC encoders, in *2012 19th IEEE International Conference on Electronics, Circuits and Systems (ICECS)* (2012), pp. 564–567.
14. G. Correa, P. Assuncao, L. Agostini, L.A. da Silva Cruz, Motion compensated tree depth limitation for complexity control of HEVC encoding, in *2012 IEEE International Conference on Image Processing* (2012), pp. 217–220.
15. G. Correa, P. Assuncao, L. Agostini, L.A. da Silva Cruz, Complexity control of HEVC through quadtree depth estimation, in *EUROCON, 2013 IEEE* (2013), pp. 81–86.
16. G. Correa, P. Assuncao, L. Agostini, L.A. da Silva Cruz, Coding tree depth estimation for complexity reduction of HEVC, in *2013 Data Compression Conference*, Snowbird, Utah (2013), pp. 43–52.
17. G. Correa, P. Assuncao, L.A. Da Silva Cruz, L. Agostini, Constrained encoding structures for computational complexity scalability in HEVC, in *2013 Picture Coding Symposium*, San Jose, USA (2013).
18. G. Correa, P. Assuncao, L. Agostini, L. Silva Cruz, Complexity scalability for real-time HEVC encoders, J. Real-Time Image Process. 1–16 (2014) http://link.springer.com/article/10.1007/s11554-013-0392-8
19. G. Correa, P. Assuncao, L. Agostini, L.A. da Silva Cruz, Classification-based early termination for coding tree structure decision in HEVC, in *IEEE International Conference on Electronics, Circuits, and Systems (ICECS 2014)*, Marseille, France, submitted: June 13 2014.
20. G. Correa, P. Assuncao, L. Agostini, L.A. da Silva Cruz, Fast HEVC encoding decisions using data mining. IEEE Trans. Circ. Syst. Video Technol **24**, 660–673 (2015)
21. G. Correa, P. Assuncao, L. Agostini, L.A. da Silva Cruz, A method for early-splitting of HEVC inter blocks based on decision trees, in *European Signal Processing Conference (EUSIPCO 2014)*, Lisbon, Portugal (2014).
22. G. Correa, P. Assuncao, L. Agostini, L.A. da Silva Cruz, Four-step algorithm for early termination in HEVC inter-frame prediction based on decision trees, in *Visual Communications and Image Processing (VCIP 2014)*, Valleta, Malta (2014).

Chapter 2
Video Coding Background

This chapter provides background information about video coding theory and practice. Initially, the basic concepts of video coding and the basic model of a hybrid video compression system are presented in this chapter. A discussion on the rate-distortion optimisation (RDO) method is then followed by a description of the high-efficiency video coding (HEVC) standard. Finally, the chapter presents an introduction to the topic of computational complexity of HEVC, which is the main focus of this book.

2.1 Basic Concepts

A digital video is a sequence of digital images to be presented sequentially to the viewer at a temporal rate high enough to ensure a smooth transition-free visual perception. In general, all individual images of a video, known as frames or pictures, have the same horizontal and vertical dimensions measured in pixels. Pixel is the name given to the numerical value of the picture elements, which are organised in matrix form.

For the purpose of encoding, a frame is usually divided into several $M \times N$ blocks of pixels, where M is the number of rows and N is the number of columns. Each video coding standard defines a different block size or even a variable range for it. Usually, the modules that compose a video encoder operate using blocks of pixels as basic processing units.

The frame dimensions, also called as spatial resolution, can assume arbitrary values in different video signals, even though there are some predefined formats that are broadly used in industry, such as the common intermediate format (CIF), the video graphics array (VGA) and the 1080p, with 352×288 pixels, 640×480 pixels and 1920×1088 pixels, respectively. The higher the spatial resolution, the larger the number of pixels in it, and, consequently, the more detailed the perception of the video content.

G. Corrêa et al., *Complexity-Aware High Efficiency Video Coding*,
DOI 10.1007/978-3-319-25778-5_2

The temporal rate at which frames are presented also influences the perceived video quality. The higher the number of frames shown in a determined period, the smoother is the motion noticed and the transition from one frame to another. The number of frames per second is called the temporal resolution, sampling rate or frame rate. In general, the temporal resolution varies between 15 and 60 frames per second (fps) in typical video sequences, although higher frame rates are becoming common.

Other important aspects to be considered when encoding a video sequence are the adopted colour system and subsampling pattern. The most used colour systems are RGB (red, green and blue) and YCbCr (luminance, blue chrominance and red chrominance). Thousands of distinct colours can be perceived from the different combinations of the elements that compose these systems. In RGB, colours are formed through different combination of the R (red), G (green) and B (blue) primary components. However, as the human visual system is much more sensible to luminance than to colour information, the YCbCr system was created to take advantage of this characteristic by allowing a subsampling of the chrominance (colour) information [1].

The use of subsampling patterns is by itself a kind of video compression, since it allows simply discarding part of the video data without causing perceptible visual impacts. Several spatial subsampling configurations can be used, but the most common are the 4:2:2 and the 4:2:0. In 4:2:2, there are two blue chrominance (Cb) and two red chrominance (Cr) samples for each four luminance (Y) samples. In 4:2:0, there is only one Cb and one Cr sample for each four Y samples [2]. The 4:2:0 configuration was used in all experiments presented in this book.

2.2 Lossless and Lossy Compression

Image and video compression techniques are generally based on exploiting data redundancy and data irrelevance. Lossless compression algorithms aim at reducing redundancy in data, representing it with a smaller amount of bits without causing any loss of information. In lossless techniques for image and video compression, the compressed data can be completely recovered after the decompression process, so that the original and the reconstructed images or videos are exactly the same. On the other hand, lossy compression techniques incur in some loss of information, and the compressed data cannot be fully restored during the decompression process. Lossy algorithms for image and video compression usually exploit characteristics of the human visual system and remove data portions that are not relevant for receivers, retaining only information that can be perceived in such a way that the final, decoded image seems to have no or small difference to the original one for almost all observers.

Three basic types of redundancy can be exploited in lossless digital video compression: spatial, temporal and entropic. Spatial or intra-frame redundancy arises due to correlation between pixels in the same image. If correlation exists in the spatial domain (i.e. neighbouring pixels have similar values), redundancy can be reduced through intra-frame prediction, a technique present in most current image and video coding standards. This type of correlation is visible to the human eye, since the neighbouring pixels present similar values [3]. Temporal or inter-frame redundancy arises due to similarities between temporally adjacent frames. The values of a set of pixels in a determined region of a video might not change from one frame to another or might vary a little, as in the case of an image background (e.g. sky or wall). In other cases, the same pixels might reappear in a frame displaced in relation to a previous one, as in the case of an object moving in a scene (e.g. a bird). Efficient techniques for reducing inter-frame redundancy yield high levels of compression and for this reason all current video coding standards exploit such techniques [3]. Entropic redundancy is related to the occurrence frequency of the encoded symbols in a video. The higher the probability of a symbol occurring, the lower is the amount of information associated to it. Entropic redundancy is removed by the entropy coding module in most video coding standards [4].

Even though they provide no loss of information, redundancy reduction techniques achieve limited compression ratios. However, in many applications information loss is not exactly a problem, as long as the reconstructed data is still comprehensible by the receiver. In the specific case of digital images and video sequences, some loss of information can be tolerated by the viewer as long as it does not incur in annoying visual artefacts. In return for accepting such loss, much higher compression ratios can be achieved with lossy methods than with lossless methods.

Lossy compression is generally performed with the aid of a quantisation process, which aims at representing a large (sometimes infinite) number of possible distinct inputs as a much more limited number of code words, as in a many-to-few mapping [3]. In image and video compression, this is achieved by exploiting characteristics of the human visual system. The human eye can perceive small differences in brightness over a relatively large area (low frequency), but cannot perceive very well such variations in small areas (high frequency) [3]. This way, the amount of information that represent high-frequency components is said to be irrelevant and can be reduced in the quantisation process, which is done by simply dividing each component in the frequency domain by a constant and then rounding it to the nearest integer. As a result, many high-frequency components are rounded to small integer numbers or to zero, which yields very large compression ratios. Notice, however, that the original values cannot be restored after being rounded to the nearest integer, which characterises information loss.

2.3 Hybrid Video Compression

Hybrid video compression is based on the following signal and data processing operations: (1) inter- and intra-frame prediction, (2) de-correlating transform, (3) quantisation and (4) entropy coding. The prediction step is usually followed by the

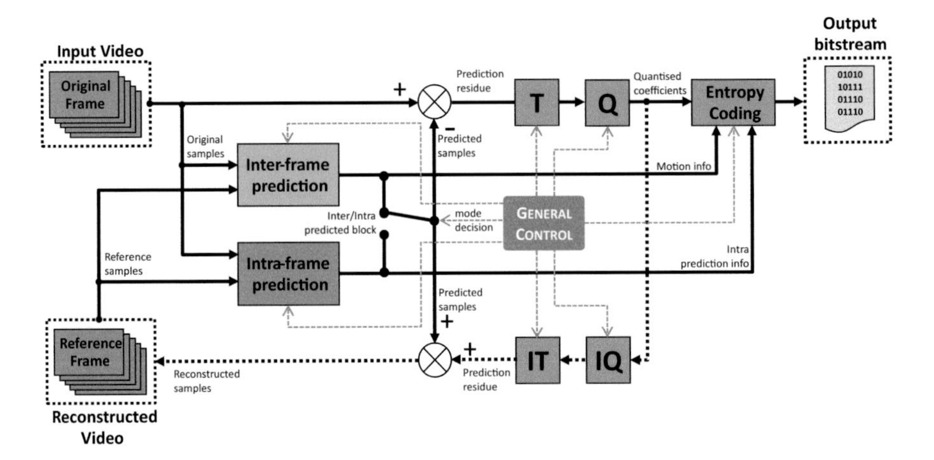

Fig. 2.1 Generic hybrid video compression system (encoder)

transform and quantisation of prediction residues, which is then followed by the entropy coding.

Figure 2.1 shows the basic components of a generic video encoder. The solid black arrows indicate the data flow at the encoder from the input video sequence to the generated output bitstream. In intra-frame prediction, each block is predicted from the pixels of reconstructed, previously encoded neighbour blocks according to a determined prediction mode. In inter-frame prediction, the block is predicted from pixels belonging to previously encoded frames.

Motion compensation (MC) is used in inter-frame prediction to exploit the fact that in most video sequences, the only difference between two adjacent frames results from camera or object motion. By using MC, the encoder is able to encode only the difference between two frames, discarding the redundant information between them. In order to better capture the scene motion occurring between adjacent frames, block-based motion estimation (ME) is usually employed to find in previously encoded frames the best matching blocks in comparison to those in the current frame. The location of the best match in the reference frame is defined by a (x, y) coordinate, called a motion vector (MV), which describes the reference block offset relatively to the current block.

In both intra- and inter-frame predictions, the predicted and the current blocks are rarely equal, so that the difference between them must be calculated, encoded and transmitted. This difference is called prediction residue or prediction error. If ME is used in inter-frame prediction, the corresponding MV has to be encoded and transmitted together with the residue (*Motion info*, in Fig. 2.1). In the case of intra-frame prediction, the mode used for prediction is encoded with the residue (*Intra prediction info*, in Fig. 2.1).

Before being encoded, the residue is processed by a mathematical transform (*T* module, in Fig. 2.1), which converts the values from the spatial domain to the frequency domain in order to de-correlate the residue and concentrate the energy in a few coefficients. Then, a quantisation step (*Q* module, in Fig. 2.1) is applied to transformed coefficients to eliminate small values associated to spectral components that are not perceptually relevant, decreasing the amount of data to be encoded without losing important information. Finally, the entropy coding processes the symbols that represent quantised transform coefficients to reduce their redundancy.

Many methods for intra-/inter-frame prediction, data transformation, quantisation and entropy coding are used in the current video coding standards. The *General Control* module presented in Fig. 2.1 is responsible for deciding which modes are tested by each module and which mode results in the best encoding performance. These decisions, shown as dashed grey lines in the figure, are performed according to the encoder implementation and affect directly its compression efficiency. A frequently used solution for this problem is the rate-distortion optimisation (RDO), which is explained in Sect. 2.5. The specific operations of the modules present in the HEVC standard are explained in Sect. 2.6.

Since the prediction signal used to compute the residue sent to the decoder must be exactly the same signal used at the decoder to reconstruct the decoded images, a reconstruction loop must be present at the encoder. Such reconstruction loop is actually a decoder operating inside the encoder, which ensures that intra-/inter-prediction at the encoder uses the same sample values as the decoder. The data flow in the reconstruction loop is presented with dashed black arrows in Fig. 2.1.

The decoder is presented in Fig. 2.2. The input bitstream is parsed and decoded by the entropy decoder, and information such as MVs, reference frame indices, intra-prediction mode and coding mode are sent to their respective modules. The inverse quantisation (*IQ* module, in Fig. 2.2) and the inverse transform (*IT* module, in Fig. 2.2) process the quantised and transformed coefficients, respectively, generating the prediction residue that is added to the predicted samples obtained from the inter-frame or intra-frame decoding modules. Finally, the reconstructed samples (i.e. the residue added to the predicted samples) are stored and used as references by the intra-frame or the inter-frame prediction modules.

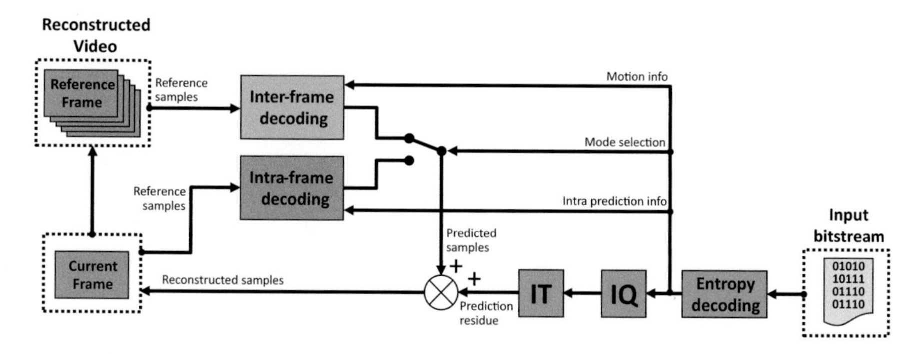

Fig. 2.2 Generic hybrid video compression system (decoder)

2.4 Distortion Metrics

Since all video coding standards introduce distortion, it is important to review the objective quality measures commonly used in video coding. Quality is a very difficult parameter to define and evaluate, because it depends on the viewer and such subjective nature of video quality is not easy to measure objectively. Although there are computable quality measures that estimate subjective grades, they are not used in the experiments described in this book because there is no wide consensus yet on which ones. Objective metrics, on the other hand, are based on direct comparisons between pixels in two different images or blocks, such as the original and the reconstructed image, and for this reason they are easily computable and widely understood even in regard to their limitations.

The most used and known objective distortion metric is the peak signal-to-noise ratio (PSNR) [2], defined in Eq. (2.1), where MAX is the maximum value that a sample can assume (2^{n-1}, where n is the number of bits per sample) and MSE is the mean-squared error (MSE) for the image or block, calculated as in Eq. (2.2). In Eq. (2.2), m and n are the image dimensions, and O and R represent the original and reconstructed luminance or chrominance samples, respectively. MSE is by itself a distortion metric that quantifies the difference between the samples from two images or blocks. Just as MSE, there are other metrics with the same purpose, which perform this function with larger or smaller computational complexity. This is the case of the sum of absolute differences (SAD), a low-complexity distortion metric broadly used in video encoders. SAD is computed as shown in Eq. (2.3), where O, R, m and n assume the same meaning as in Eq. (2.2):

$$\mathrm{PSNR}_{\mathrm{dB}} = 20 \cdot \log_{10}\left(\frac{\mathrm{MAX}}{\sqrt{\mathrm{MSE}}}\right) \tag{2.1}$$

$$\mathrm{MSE} = \frac{1}{mn}\sum_{i=0}^{m-1}\sum_{j=0}^{n-1}\left(R_{i,j} - O_{i,j}\right)^2 \tag{2.2}$$

$$\mathrm{SAD} = \sum_{i=0}^{m-1}\sum_{j=0}^{n-1}\left|R_{i,j} - O_{i,j}\right| \tag{2.3}$$

2.4.1 Bjøntegaard Model

Based on the PSNR distortion metric, Gisle Bjøntegaard proposed in [5] a model that measures the compression efficiency difference between two algorithms. In general terms, the Bjøntegaard delta PSNR (BD-PSNR) measure corresponds to the average PSNR difference in decibels (dB) for two different encoding algorithms considering the same bit rate. Similarly, the Bjøntegaard delta rate (BD-rate) reports the average bit rate difference in percent for two different encoding algorithms considering the same PSNR.

A third-order logarithmic polynomial fitting is used to approximate a rate-distortion (R-D) curve given by a set of N bit rate values ($R_1, \ldots, R_N$) with their corresponding PSNR measurements ($D_1, \ldots, D_N$), as in Eq. (2.4), where $\breve{D}(R)$ is the fitted distortion in PSNR, R is the output bit rate and a, b, c and d are fitting parameters:

$$\breve{D}(R) = a \cdot \log^3 \cdot R + b \cdot \log^2 \cdot R + c \cdot \log \cdot R + d \tag{2.4}$$

To simplify notation, Eq. (2.4) is rewritten as Eq. (2.5), considering r as the logarithm of the bit rate (i.e. $r = \log R$):

$$\breve{D}(r) = a \cdot r^3 + b \cdot r^2 + c \cdot r + d \tag{2.5}$$

With four R-D pairs (i.e. four bit rate values and their corresponding PSNR measurements) obtained from actual encodings, the four fitting parameters can be solved for a curve. Then, the average PSNR difference between two R-D curves corresponding to two different algorithms can be approximated by the difference between the integrals of the fitted curves divided by the integration interval, as in Eq. (2.6), where ΔD is the BD-PSNR between the two fitted R-D curves $\breve{D}_1(r)$ and $\breve{D}_2(r)$ and r_L and r_H are the integration bounds obtained as in Eqs. (2.7) and (2.8), respectively [5]. In Eqs. (2.7) and (2.8), $r_{x,y}$ represents the bit rate value of point y belonging to curve $\breve{D}_x(r)$, and N_1 and N_2 are the number of points in each curve:

$$D = \frac{\int_{r_L}^{r_H} \left(\breve{D}_2(r) - \breve{D}_1(r) \right) dr}{r_H - r_L} \tag{2.6}$$

$$r_L = \max\left\{ \min\left(r_{1,1}, \ldots, r_{1,N_1} \right), \min\left(r_{2,1}, \ldots, r_{2,N_2} \right) \right\} \tag{2.7}$$

$$r_H = \min\left\{ \max\left(r_{1,1}, \ldots, r_{1,N_1} \right), \max\left(r_{2,1}, \ldots, r_{2,N_2} \right) \right\} \tag{2.8}$$

The bit rate as a function of the distortion is also expressed in the Bjøntegaard model through the third-order polynomial given in Eq. (2.9). The average bit rate difference between two R-D curves is approximated as in Eq. (2.10), where ΔR is the BD-rate between the two fitted R-D curves $\breve{r}_1(D)$ and $\breve{r}_2(D)$ and D_L and D_H are the integration bounds obtained as in Eqs. (2.11) and (2.12), respectively [5]. In Eqs. (2.11) and (2.12), $D_{x,y}$ represents the distortion value of point y belonging to curve $\breve{r}_x(D)$, and N_1 and N_2 are the number of points in each curve:

$$\breve{r}(D) = a \cdot D^3 + b \cdot D^2 + c \cdot D + d \tag{2.9}$$

$$R = 10^{\frac{1}{D_{\mathrm{H}} - D_{\mathrm{L}}} \int_{D_{\mathrm{L}}}^{D_{\mathrm{H}}} \left[\breve{r}2(D) - \breve{r}1(D) \right] \mathrm{d}D} - 1 \tag{2.10}$$

$$D_{\mathrm{L}} = \max \left\{ \min \left(D_{1,1},, \ldots,, D_{1,N_1} \right), \min \left(D_{2,1},, \ldots,, D_{2,N_2} \right) \right\} \tag{2.11}$$

$$D_{\mathrm{H}} = \min \left\{ \max \left(D_{1,1},, \ldots,, D_{1,N_1} \right), \max \left(D_{2,1},, \ldots,, D_{2,N_2} \right) \right\} \tag{2.12}$$

All methods proposed in this book for complexity reduction and scaling are compared to the original encoder version and to related works using the Bjøntegaard measures.

2.5 Rate-Distortion Optimisation

In order to achieve optimal R-D efficiency, a video encoder must be able to select the best coding modes for any particular video sequence. In other words, given a particular bit rate constraint, the encoder must select a coding mode that returns minimal image distortion. This constrained optimisation problem can be mathematically described as follows.

Let S represent all allowable modes and i represent an element of S (i.e. $i \in S$). Then,

$$\begin{gathered} i^* = \arg \min_{i \in S} D(i) \\ \text{subject to} \;\; R(i) \leq R_{\mathrm{T}} \end{gathered}, \tag{2.13}$$

where i^* is the optimal mode that minimises the distortion, $D(i)$ is the distortion obtained with mode i, $R(i)$ is the number of bits obtained with mode i and R_{T} is the bit rate constraint.

This optimisation problem can be solved using Lagrangian methods, in which the distortion term is weighted against the bit rate term, giving rise to an unconstrained problem [6]. The Lagrangian minimisation is represented in Eq. (2.14), where $J(i)$ is the R-D cost, $D(i)$ and $R(i)$ are the resulting distortion and bit rate when using mode i and λ is the Lagrangian multiplier [7, 8]. The coding mode i^* that returns the minimum cost $J(i)$ is selected as the solution of the following unconstrained problem:

$$\begin{gathered} i^* = \arg \min_{i \in S} J(i) \\ \text{where} \;\; J(i) = D(i) + \lambda \cdot R(i) \end{gathered} \tag{2.14}$$

As there are no simple models to describe the relationship between the coding modes and the R-D cost J, the RDO process implemented in current video encoders tests all possible coding modes and selects the one that results in the smallest R-D

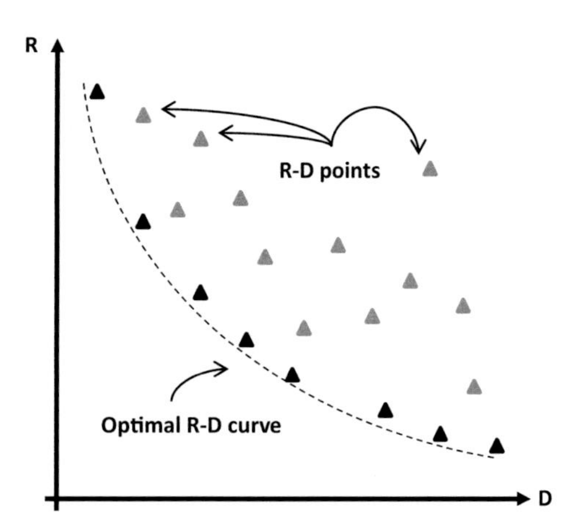

Fig. 2.3 Rate-distortion (R-D) points and optimal curve

cost. Figure 2.3 shows an example of an R-D space and an optimal R-D curve. In the figure, each point corresponds to a different mode tested, and those that present the smallest distortion (*D* axis) for a given target bit rate (*R* axis) are in the optimal R-D curve.

The RDO technique is very effective and yields the best possible choice among all the encoding parameter sets. However, if every combination of operating modes is tested and evaluated by exhaustive search over the parameter and mode space, the computational complexity becomes a limiting factor. In practical applications, a number of modes must be ignored in the RDO process to comply with the limitations imposed by available computational resources, but the selection of such modes is not a trivial task. Understanding the operations of video coding standards is the key to design a video coding system that manages efficiently the computational resources consumed by the encoding process.

2.6 High-Efficiency Video Coding

The high-efficiency video coding (HEVC) standard was finalised in March of 2013 by the Joint Collaborative Team on Video Coding (JCT-VC), a joint project of the ITU-T Video Coding Experts Group (VCEG) and the ISO/IEC Moving Picture Experts Group (MPEG). The standard has been designed to address the growing popularity of high-definition (HD) video and the emergence of beyond HD formats, with support to encode multiple video views [9]. Currently, the high-efficiency video coding (HEVC) standard is gradually being introduced in the market and is expected to substitute H.264/AVC during the next years as the state-of-the-art video coding standard. This new standard is able to reduce bit rates by about 40–50 % over H.264/AVC through the use of several new tools and operating modes [7]. Obviously, as more operating modes are made available, the computational complexity involved in the encoding process of HEVC becomes higher than that of previous standards, as shown later in this book.

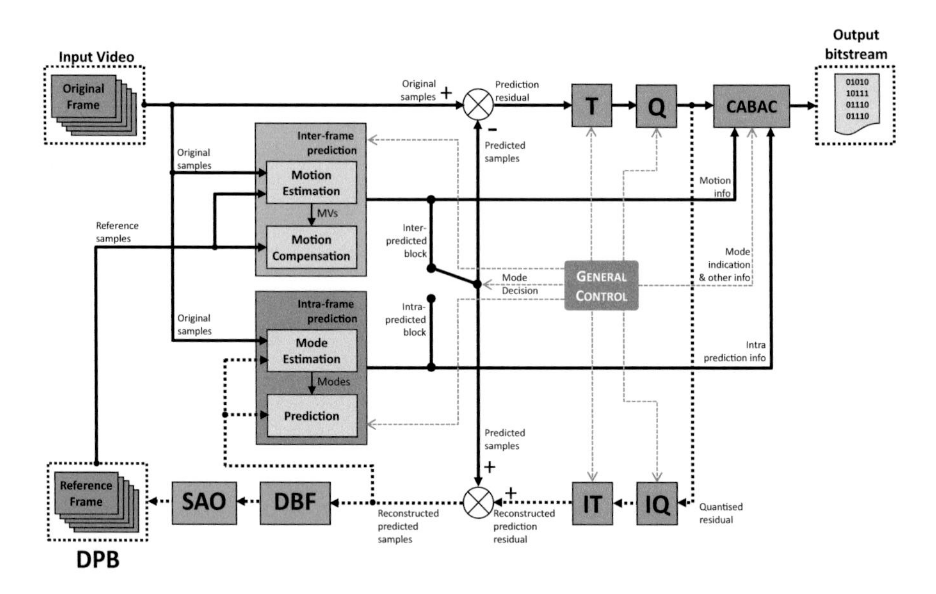

Fig. 2.4 Block diagram of a typical HEVC encoder

Figure 2.4 presents the block diagram for the HEVC encoder. Initially, each frame of the input video is split into equal-sized block-shaped regions. The first frame in the video sequence is encoded using only intra-frame prediction, since there are no frames previously encoded to be used as reference in inter-frame prediction. The remaining frames may use both intra- and inter-frame predictions.

During inter-frame prediction, the encoder predicts each block from previously encoded frames by using ME and MC. MVs obtained from the ME process determine the relative location of the best prediction block in the reference frame, which are used in the MC process. In intra-frame prediction, the *Prediction* module in Fig. 2.4 uses samples from the original and neighbouring reconstructed blocks from the current frame to predict a block according to a particular prediction mode, which is determined by the *Mode Estimation* module in Fig. 2.4.

After computing the prediction, the encoder calculates the residual, which is transformed by a linear spatial transform (*T* module, in Fig. 2.4), quantised (*Q* module, in Fig. 2.4) and entropy coded (*CABAC* module, in Fig. 2.4), generating bits that are multiplexed with motion and intra-prediction information, mode indication and other encoding information, finally resulting in the compressed bitstream.

Similar to the generic hybrid video encoder presented in Fig. 2.1, the HEVC encoder also duplicates the decoder processing loop, such that both the decoder and the encoder can use the same reference samples for intra-/inter-frame prediction. Therefore, the quantised residue is fed to the inverse quantisation and inverse transform (*IQ* and *IT* modules, in Fig. 2.4) in order to reconstruct the residual information, which is added to the predicted samples to generate the reconstructed samples. The result of the addition is delivered to the deblocking filter (*DBF* module, in Fig. 2.4) and the sample adaptive offset (*SAO* module, in Fig. 2.4) to smooth out artefacts caused by block-wise processing and quantisation. The filtered, reconstructed samples are finally stored in the decoded picture buffer (*DPB*, in Fig. 2.4) and used for prediction in future frames. The decoding loop is presented with dashed black arrows in Fig. 2.4.

After this brief overview of the HEVC encoding process, the next subsections explain in more detail the new features introduced by the standard.

2.6.1 Encoding Structures

Even though HEVC is also based on the classic block-based video coding scheme of previous standards, significant modifications were introduced to its encoding data structures, especially regarding frame partitioning. The authors' contribution presented later in this book relies heavily on methods for deciding the best frame partitioning structures during the encoding process, so that this subsection provides most of the necessary background for the comprehension of the next chapters.

2.6.1.1 Video Partitioning Structures

In HEVC, each video sequence is divided into groups of pictures (GOPs), which are limited by two frames that constitute random access points (RAPs) from which the decoder can start decoding without needing any previous frames. Figure 2.5 illustrates a video sequence divided into a number of GOPs, where the frames in black represent RAPs.

A GOP is composed of a set of pictures, or frames, which may be divided into a set of slices. Each slice is a part of the frame that can be decoded independently from other slices in the same frame, which is very useful in case of data losses or when applying parallel processing strategies, as shown later. Figure 2.6 illustrates a frame divided into two slices.

A slice is composed of sequential coding tree units (CTU), which are further divided into coding units (CUs). These structures are explained in detail in the next section. All CUs inside each slice are encoded according to the slice type, which

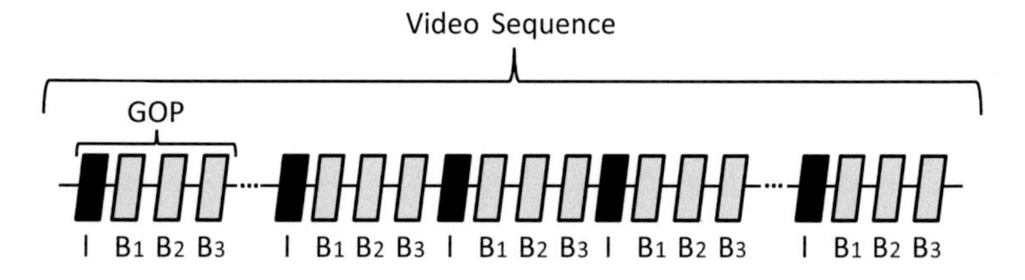

Fig. 2.5 Video sequence divided into a number of GOPs

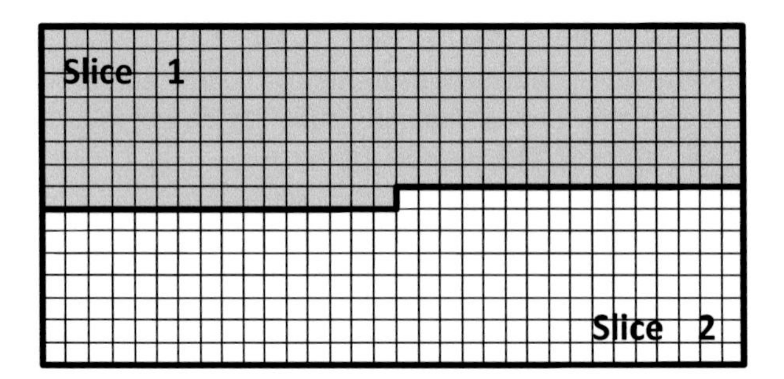

Fig. 2.6 Frame divided into two slices and several CTUs

may be I, P or B. In I slices, all CUs are encoded using only the intra-frame prediction. In a P slice, besides the intra-frame prediction used in I slices, the CUs can also be encoded using inter-frame prediction with one reference for MC (i.e. unidirectional prediction). In B slices, besides intra-frame prediction and unidirectional inter-frame prediction, CUs can also be encoded using inter-frame prediction with two references for MC (i.e. bidirectional prediction). Further details are presented in Sect. 2.6.2.2.

2.6.1.2 Frame Partitioning Structures

A slice is partitioned into a number of square blocks of equal size called coding tree units (CTU), as shown in Fig. 2.6. Considering the 4:2:0 subsampling configuration, each CTU is composed of one luminance coding tree block (CTB) of size $W \times W$ and two chrominance CTBs of size $(W/2) \times (W/2)$, where W may be equal to 8, 16, 32 or 64. The luminance CTB and the two chrominance CTBs form the CTU, which is considered the basic processing unit of HEVC.

Each CTB in a CTU can be divided into smaller blocks, called coding blocks (CB), in the form of a quadtree structure, called the coding tree. The CTB is the root of this quadtree, which may assume variable depth, according to the encoder configuration. The largest coding block (LCB) size and the smallest coding block (SCB) size used by the encoder are defined in its configuration, but the maximum and minimum allowed sizes for a CB in HEVC are 64×64 and 8×8, respectively, so that up to four coding tree depths are possible. The same tree structure is usually applied to luminance and chrominance CTBs, except when the minimum size for chrominance blocks is reached. In the HEVC model (HM) encoder [10], the coding tree structure is defined through an iterative splitting process, which evaluates all possibilities in an RDO-based scheme, until the minimum CB size, which is usually 8×8 for luminance CTBs, is reached.

Figure 2.7 shows an example of a 64×64 CTB divided into several CBs. The example shows the final coding tree division chosen after all the possibilities are evaluated. The tree leaves (grey blocks) are the final CBs belonging to the encoded

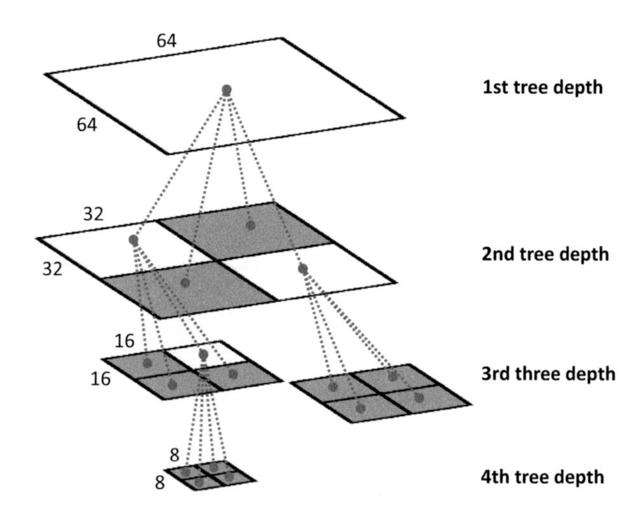

Fig. 2.7 Coding tree structure of a CTB divided into CBs

quadtree. This flexible encoding structure allows the use of large CBs to encode large homogenous regions of a frame and small CBs in regions with more detailed texture.

For intra- and inter-frame prediction, each CB may be divided into two or four prediction blocks (PBs), which are separately predicted. All PBs in a CB are predicted with either inter-frame or intra-frame prediction, so that the CB is said to be an inter-CB or an intra-CB. Figure 2.8 presents all possible PB splittings that can be used for a CB. These possibilities are called PB splitting modes from now on in this book in order to distinguish them from the prediction modes, which will be defined later in this chapter. In HM [10], the best PB splitting mode is chosen with recourse to an RDO-based scheme, which evaluates the prediction using all PB splitting mode possibilities and compares their use in terms of bit rate and distortion. The total number of possibilities varies according to the CB size, as specified in Table 2.1. The same PB splitting mode is used for luminance and chrominance PBs, which together form the prediction unit (PU).

When transform coding the prediction residual, each CB is assumed to be the root of another quadtree-based structure called residual quadtree (RQT). Likewise the CTBs, each CB is recursively partitioned into transform blocks (TB), which are the basic units to which both transform and quantisation operations are applied. The leaves of the RQT (grey nodes, in Fig. 2.9) are the final TBs, which are chosen in an RDO-based scheme in the HM encoder. The maximum and minimum TB sizes used by the encoder are defined in its configuration, but the maximum allowed size for a TB is 32×32 and the minimum size is 4×4, so that up to four RQT depths are possible. The same RQT structure is used for both luminance and chrominance CBs. A transform unit (TU) is formed by the luminance TB and its two associated chrominance TBs.

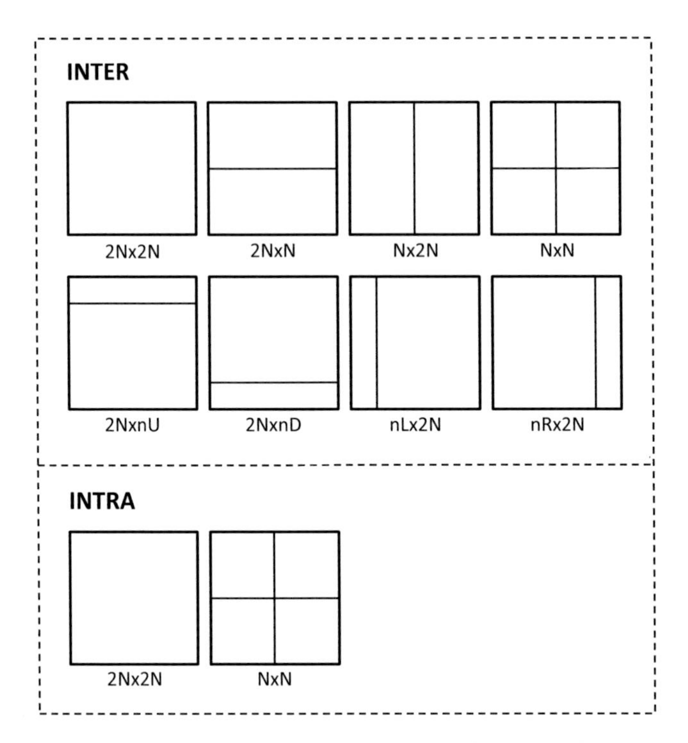

Fig. 2.8 Inter-frame and intra-frame PB splitting modes available in HEVC

Table 2.1 PB splitting modes available for each CB size in HEVC

	PB splitting modes	
CB size	Inter	Intra
Larger than SCB	$2N\times 2N$, $2N\times N$, $N\times 2N$, $2N\times nU$, $2N\times nD$, $nL\times 2N$, $nR\times 2N$	$2N\times 2N$
SCB	$2N\times 2N$, $2N\times N$, $N\times 2N$, $N\times N$	$2N\times 2N$, $N\times N$

2.6.1.3 Parallel Processing Structures

The HEVC standard defines three data partitioning and processing orders designed to facilitate parallel processing. The use of these features may be decided depending on the encoder application context.

Tiles are defined as rectangular image regions, which enable a coarse but easy-to-implement parallel processing without requiring sophisticated thread synchronisation. Tiles can be independently decoded and encoded sharing some header information. When using tiles, the encoder segments the frame into rectangular regions, as shown in the example of Fig. 2.10, encodes and transmits them in raster scan order. All data dependencies are broken at the tile boundaries, so that independent encoding is performed for each one of them. The only exception is the deblock-

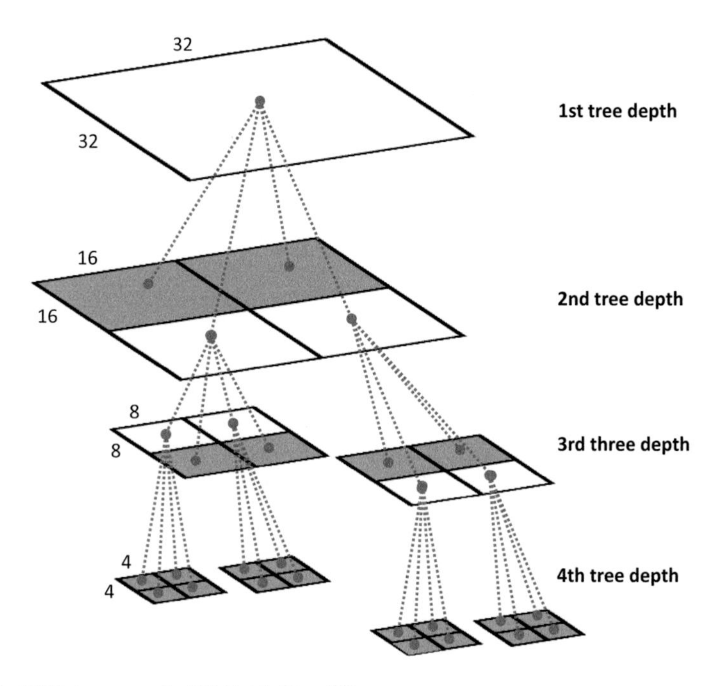

Fig. 2.9 RQT structure of a CB divided into TBs

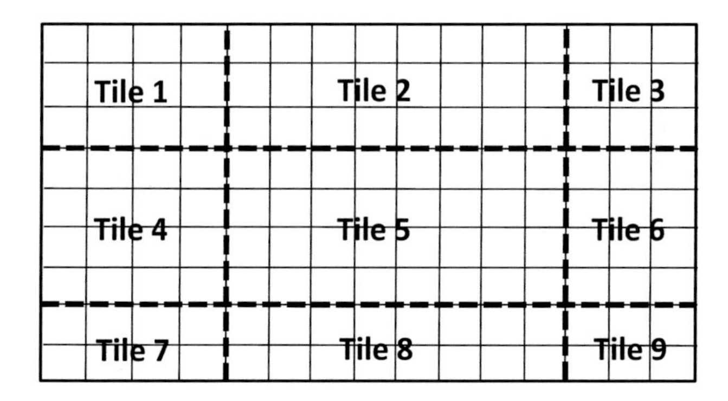

Fig. 2.10 A frame divided into nine tiles

ing filter (DBF), which can be applied across tile boundaries to reduce visual artefacts.

By using wavefront parallel processing (WPP), a much finer degree of parallelism can be achieved. The slice is divided into rows of CTUs, which are processed according to the order presented in Fig. 2.11. The first CTU row of each slice is encoded in an ordinary way. The second CTU row starts being encoded after the two first CTUs in the first row are encoded, the third CTU row starts being encoded after the two first CTUs in the second row are encoded, and so on. The main advantage

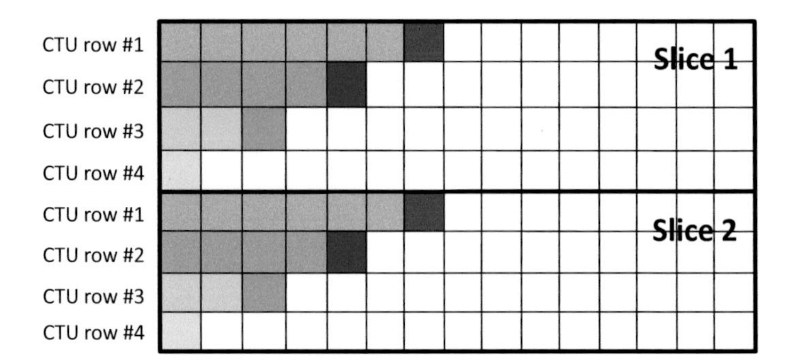

Fig. 2.11 Wavefront parallel processing encoding order

in the use of WPP instead of tiles is the possibility of performing inter-frame or intra-frame prediction across the WPP boundaries, increasing the encoder compression efficiency.

Dependent slice segments allow data associated with a determined WPP or tile to be encoded and assembled in a separate logical data packet for transmission, making that data available for fragmented packetisation with lower latency than if it were all coded together in one slice.

2.6.2 *Encoding Process*

This section presents a description of the modules that compose the HEVC encoder (Fig. 2.4). As the intra-/inter-frame prediction, transform (*T*) and quantisation (*Q*) modules are the most important for the comprehension of the methods presented later in this book, they are described here in more detail, while a briefer overview is presented for the remaining modules.

2.6.2.1 Intra-frame Prediction

As previously explained, the intra-frame prediction is responsible for decreasing spatial redundancy by predicting the samples of the current PB from those of neighbouring PBs already encoded in the same slice, i.e. located at the left and above the current PB. As previously shown in Fig. 2.8, a CB can be split according to two different splitting modes for intra-prediction: $2N\times 2N$ and $N\times N$. Figure 2.12 shows an example of intra-frame prediction in an $N\times N$ PB of size 16×16, where the samples within the current PB are presented in white, the left neighbouring samples in light grey, the top neighbouring samples in dark grey and the top-left diagonal sample in black. In total, $4N+1$ samples are used for intra-prediction in an $N\times N$ PB. If the CB has dimensions equal to the SCB, both $2N\times 2N$ and $N\times N$ splitting modes are available for intra-prediction. Otherwise, only the $2N\times 2N$ splitting mode is available.

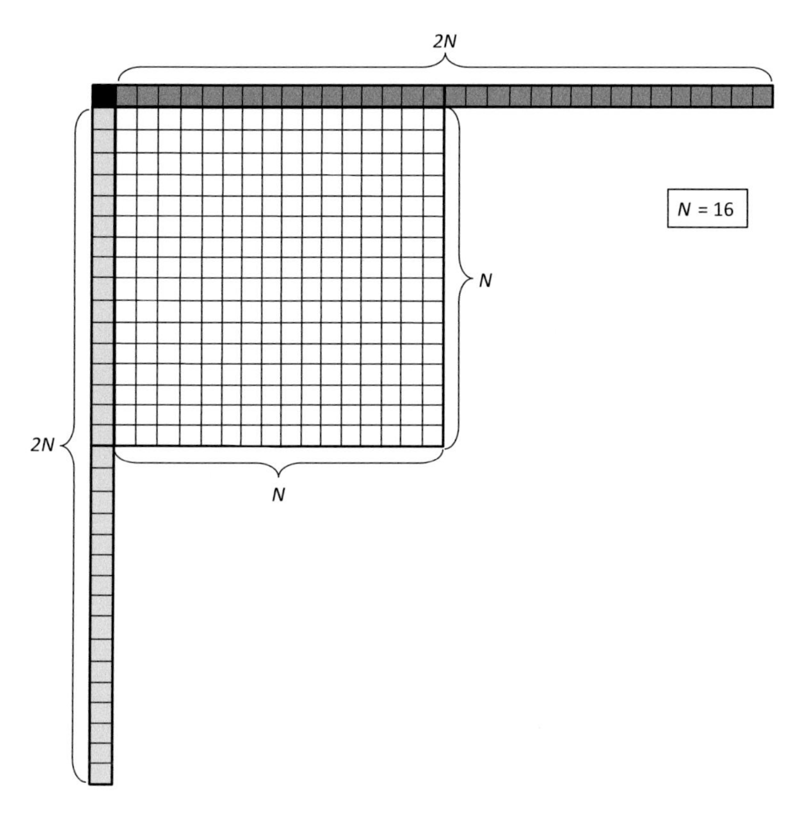

Fig. 2.12 Neighbouring samples used for intra-prediction in an $N \times N$ PB of size 16×16

The intra-frame prediction of HEVC is very similar to that of H.264/AVC and is essentially extended to allow more prediction modes. As Fig. 2.13 shows, HEVC supports a total of 33 angular modes, a DC (flat) mode and a planar (surface fitting) mode [9]. When one of the angular modes is used, the PB is predicted directionally from spatially neighbouring samples that were previously reconstructed, but not yet filtered, as shown in Fig. 2.4 by the input arrows of the intra-frame prediction module. The angular prediction consists of simply copying the neighbouring samples to the predicted block [11]. The DC mode computes an average of the neighbouring reference samples and uses it for all samples within the PB, building a flat surface for the whole block. Alternatively, the planar mode calculates average values of two linear predictions using four corner reference samples, building an amplitude surface with a horizontal and vertical slope derived from the boundaries [9].

In order to decrease the computational complexity of the intra-frame prediction process, the HM reference software implements a rough mode decision (RMD), or intra-prediction mode estimation, which is followed by the actual prediction with pixel estimate computations. The RMD method was proposed in [12] and incorporated into the HM encoder as a solution to decrease the number of R-D cost computations for the intra-mode decision. It consists in constructing a candidate mode list

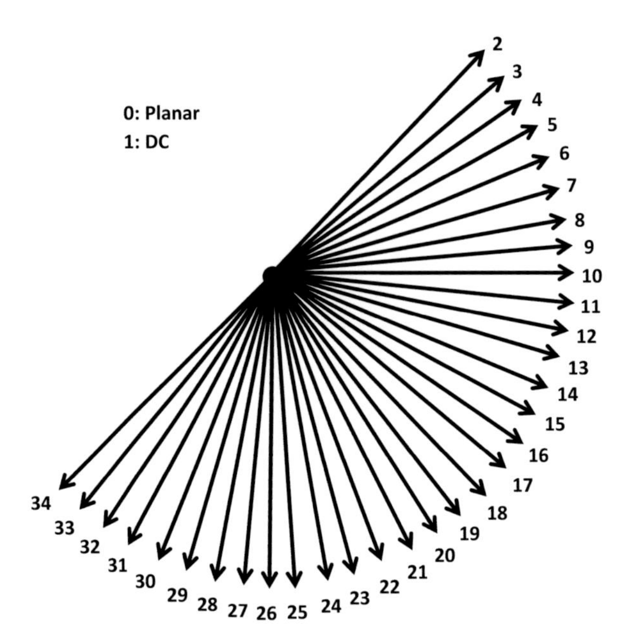

Fig. 2.13 Intra-prediction modes tested for each PB

with the three best modes for 64×64, 32×32 and 16×16 PBs and the eight best modes in the case of 8×8 and 4×4 PBs.

The best modes are those which resulted in the smallest low-complexity R-D cost (J_{LRD}), computed as in Eq. (2.15), where *SATD* is the sum of absolute values of the Hadamard transformed coefficients of the prediction residue, λ_{PRED} is the Lagrangian multiplier and R_{PRED} is the number of bits necessary to encode the prediction mode information. Besides the best modes determined in this process, the list of candidate modes is also composed of the most probable modes (MPM), which are inferred from the neighbouring PBs.

$$J_{\text{LRD}} = \text{SATD} + \lambda_{\text{PRED}} \cdot R_{\text{PRED}} \tag{2.15}$$

After defining the list of candidate modes, the full R-D cost (J_{FRD}) for each of these modes is computed using Eq. (2.16) [12], where SSE_{Luma} and SSE_{Chroma} are the sums of squared errors between the original block and the predicted block for luminance and chrominance signals, respectively. λ_{MODE} is the Lagrangian multiplier and R_{MODE} is the number of bits to encode the candidate mode. The mode with the smallest J_{FRD} is then used in the actual prediction, which precedes the residue computation and transform coding.

$$J_{\text{FRD}} = \left(\text{SSE}_{\text{Luma}} + 0.57 \cdot \text{SSE}_{\text{Chroma}}\right) + \lambda_{\text{MODE}} \cdot R_{\text{MODE}} \tag{2.16}$$

The Lagrangian multipliers λ_{PRED} and λ_{MODE} in Eqs. (2.15) and (2.16), respectively, are calculated in the HM encoder as a function of the QP, the number of reference frames and the temporal encoding configuration [10].

By using the RMD method, the computational complexity of the overall intra-frame prediction is decreased because the full R-D is computed only for those modes selected for the candidate list. However, the prediction step is still performed for every possible mode for a PB in order to calculate the low-complexity R-D costs.

2.6.2.2 Inter-frame Prediction

The inter-frame prediction of HEVC is based on MC, which allows predicting a PB using equal-sized areas from previously encoded frames used as references. The MC process forms a block of shifted pixels from the referenced frame by using a motion vector (MV), which describes the displacement of the PB from the reference to the current frame. The predicted block is then subtracted from the current block to compute the prediction residues that are inputted to the T and Q modules.

The process of determining the MV for a PB is the ME, which is the most demanding operation of the HEVC encoder in terms of computational complexity, as Chap. 4 will show. The role of ME is to search within the reference frames for the most similar region to the current PB. When the best matching region is found, the ME algorithm computes the corresponding MV with a vertical and a horizontal component, indicating the relative location of that region with respect to the current PB position.

Similar to H.264/AVC, HEVC supports ME with one quarter of pixel accuracy for luminance samples and one eighth of pixel accuracy for chrominance samples in the case of 4:2:0 format. In order to obtain the fractional position luminance samples, two separable one-dimensional filters (one 8-tap filter and one 7-tap filter) are applied horizontally and vertically to generate the half-pixel and quarter-pixel samples, respectively. Figure 2.14 shows two examples of fractional position samples (a and b, in black) that were generated based on the integer position samples (in grey) located in the same line or column (dashed rectangles).

In the example given in Fig. 2.14, samples a and b are calculated as in Eqs. (2.17) and (2.18), respectively, where the >> operator denotes arithmetic right shift, B is the bit depth of the reference samples (usually $B=8$) and *qfilter* is the vector with filter coefficients for quarter-pixel interpolation, given in Table 2.2. Table 2.2 also presents the filter coefficients for half-pixel interpolation (*hfilter*) [9].

Fractional chrominance samples are calculated similar to the luminance samples by applying one-dimensional 4-tap interpolation filters. A set of four 4-tap filters are available for chrominance interpolation, and the applied filter is chosen according to the distance between the fractional and the integer pixel. More details on luminance and chrominance interpolation filters can be found in [9].

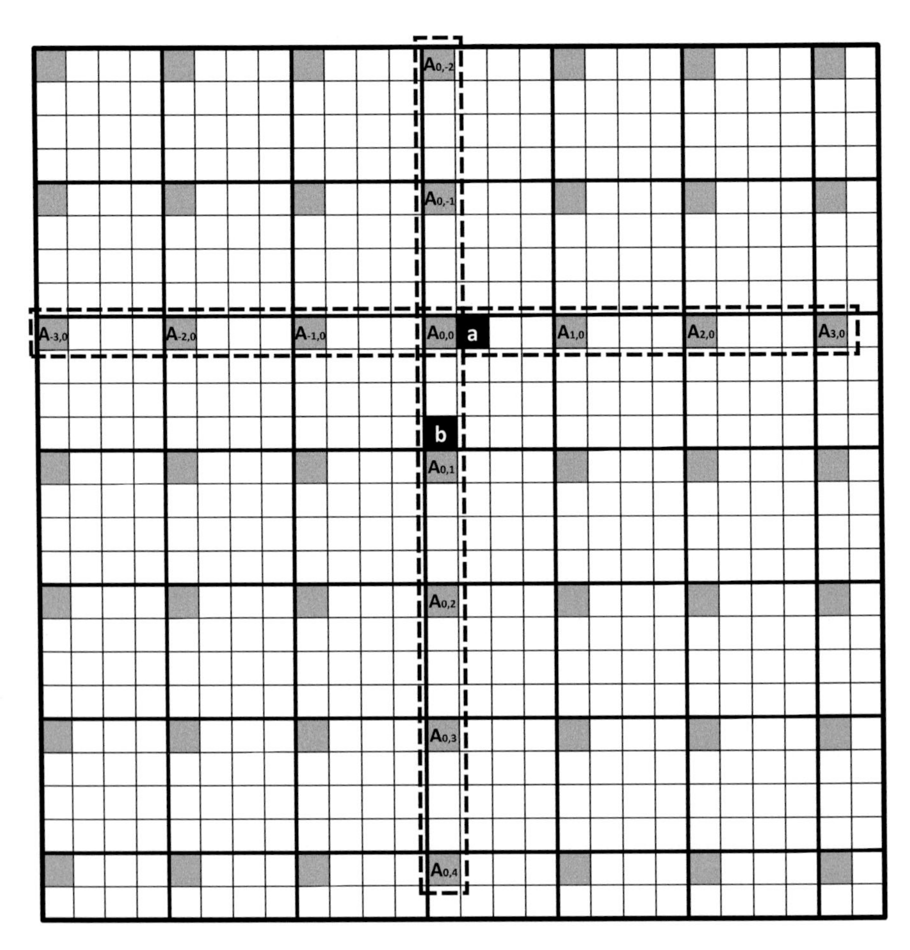

Fig. 2.14 Examples of subpixel interpolations from full-pixel samples

Table 2.2 Filter coefficients for half-pixel and quarter-pixel luminance sample interpolation

index	−3	−2	−1	0	1	2	3	4
hfilter	−1	4	−11	40	40	−11	4	1
qfilter	−1	4	−10	58	17	−5	1	

$$a = \left(\sum_{i=-3}^{3} A_{i,0} \cdot \text{qfilter}[i] \right) \gg (B-8) \tag{2.17}$$

$$b = \left(\sum_{j=-2}^{4} A_{0,j} \cdot \text{qfilter}[1-j] \right) \gg (B-8) \tag{2.18}$$

Another important aspect of the ME process is the use of multiple reference frames. The encoder maintains two lists (*List 0* and *List 1*) of reconstructed images to be used as references in the next frames. *List 0* contains the indices of past frames in display order and *List 1* contains the indices to future frames encoded out of order, as explained in Sect. 2.6.3.1. P slices allow the use of only one reference picture indexed by *List 0*, and B slices can use up to two references indexed by both *List 0* and *List 1*. When bi-prediction is used in B slices, the encoder computes an average of the prediction performed from *List 0* and the prediction performed from *List 1*. Figure 2.15 presents an example of four PBs in a frame, which are predicted from multiple references.

Block-matching ME is used to determine the actual MVs representing the displacement of PBs in the current frame in relation to the best matching area in the reference frame. Several algorithms have been proposed for searching candidate blocks, and the algorithmically simplest but the most demanding in terms of computation cost is the full search (FS). In FS, the best match is found by searching all possible candidate blocks in the search area (SA), which leads to the optimal result. However, the computational complexity involved in this process is very high and faster approaches need to be used.

Several fast motion estimation (FME) techniques have been proposed in the literature in order to decrease the number of candidate blocks in the SA. Examples of algorithms implemented in the H.264/AVC reference software are the unsymmetrical-cross multi-hexagon-grid search (UMHexS), the simple UMHexS (SUMHexS) and the enhanced predictive zonal search (EPZS) [13]. All of them are suboptimal algorithms with R-D performance equal to or smaller than that obtained with FS, with the advantage of significantly decreasing the ME computational complexity. The HM encoder uses both the FS and the suboptimal test zone search (TZS) algorithm [14] for ME with a fixed search range that can be configured before the encoding process starts.

Besides the MC-/ME-based prediction described in the previous paragraphs, the inter-frame prediction of HEVC allows the use of a merge/SKIP mode (MSM) [15],

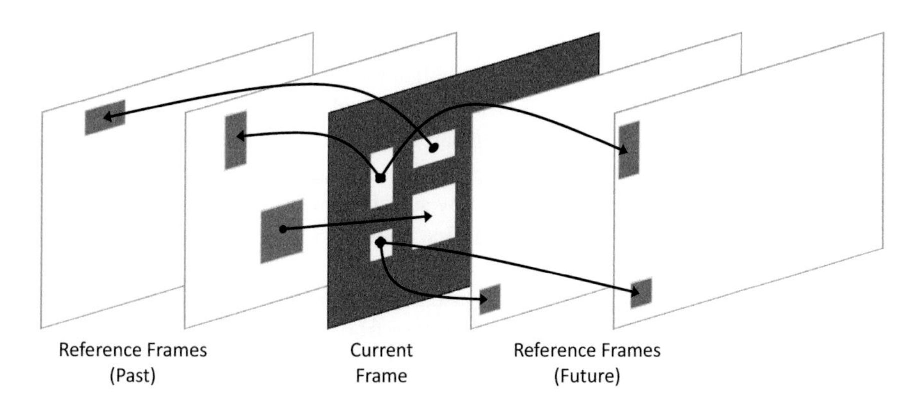

Fig. 2.15 Multiple reference frame prediction

which is conceptually similar to the *SKIP* mode in H.264/AVC. With MSM, the encoder can derive the motion information (i.e. the MV, one or two reference picture indices and the list associated to each index) from spatially or temporally neighbouring blocks, forming a merged region that shares the same MVs and picture indices. When MSM is selected for a CB, the encoder sends an index for a list containing all spatial and temporal neighbouring PBs available, known as the merge list. This index identifies which neighbouring PB is to be used as the source of motion information for the current PB. Besides the merge list index, the encoder also sends the reference picture list number and reference picture index to which the neighbour PB belongs. The *SKIP* mode also exists in HEVC, but it is treated as a special case of MSM. The *SKIP* mode is used when all coded block flags (CBF) are equal to zero in a determined CB. In this case, only a *SKIP* flag and its corresponding merge index are transmitted to the decoder. The CBF will be explained in Sect. 2.6.2.3.

When MSM is not used, the MV is encoded through the use of a motion vector prediction (MVP) method [16]. Similar to MSM, the encoder chooses an MV predictor among a set of multiple candidates derived from spatially neighbouring blocks. Then, the difference between the actual MV and the MV predictor, called differential motion vector (DMV), is transmitted to the decoder together with the index of the candidate MV predictor chosen from the MVP list. When the reference index of the neighbouring PB is not the same as the current PB, the MV is scaled according to the temporal distance between the current and the reference pictures.

2.6.2.3 Transform and Quantisation

The transform module (*T* block, in Fig. 2.4) receives the prediction residue as input and transforms it to the frequency domain. The transformed residues are then processed by the quantisation module (*Q* block, in Fig. 2.4). The HEVC standard uses a two-dimensional transform which is computed by applying one-dimensional transforms in the horizontal and vertical directions of the block. Transforms are computed for all TB tested in an RDO-based scheme.

HEVC uses integer discrete cosine transform (DCT)-based transforms of sizes 32×32, 16×16, 8×8 and 4×4, which are applied to the TB, according to its size. In HEVC, only the 32×32 transform matrix is defined and the remaining transformation matrices are derived from it by using part of its entries. For example, the transformation matrix for 16×16 TBs is presented in Eq. (2.19) [9]. The matrices for 8×8 and 4×4 TBs can be derived from it by selecting only the first eight entries of lines 0, 2, 4, 6, 8, 10, 12 and 14 and the first four entries of lines 0, 4, 8 and 12, respectively.

An alternative 4×4 integer transform based on the discrete sine transform (DST) is applied to 4×4 luma residue blocks resulting from intra-frame prediction. The reasoning behind the application of such transform is that it better fits the statistical property that residue magnitudes tend to increase as the distance from the boundary samples that are used for prediction becomes larger. In comparison to the 4×4 DCT-based transform, the use of DST-based transform to encode intra-predicted blocks results in a bit rate reduction of 1 %. The 4×4 DST-based transformation matrix is presented in Eq. (2.20) [9]:

$$H = \begin{bmatrix} 64 & 64 & 64 & 64 & 64 & 64 & 64 & 64 & 64 & 64 & 64 & 64 & 64 & 64 & 64 & 64 \\ 90 & 87 & 80 & 70 & 57 & 43 & 25 & 9 & -9 & -25 & -43 & -57 & -70 & -80 & -87 & 90 \\ 89 & 75 & 50 & 18 & -18 & -50 & -75 & -89 & -89 & -75 & -50 & -18 & 18 & 50 & 75 & 89 \\ 87 & 57 & 9 & -43 & -80 & -90 & -70 & -25 & 25 & 70 & 90 & 80 & 43 & -9 & -57 & -87 \\ 83 & 36 & -36 & -83 & -83 & -36 & 36 & 83 & 83 & 36 & -36 & -83 & -83 & -36 & 36 & 83 \\ 80 & 9 & -70 & -87 & -25 & 57 & 90 & 43 & -43 & -90 & -57 & 25 & 87 & 70 & -9 & -80 \\ 75 & -18 & -89 & -50 & 50 & 89 & 18 & -75 & -75 & 18 & 89 & 50 & -50 & -89 & -18 & 75 \\ 70 & -43 & -87 & 9 & 90 & 25 & -80 & -57 & 57 & 80 & -25 & -90 & -9 & 87 & 43 & -70 \\ 64 & -64 & -64 & 64 & 64 & -64 & -64 & 64 & 64 & -64 & -64 & 64 & 64 & -64 & -64 & 64 \\ 57 & -80 & -25 & 90 & -9 & -87 & 43 & 70 & -70 & -43 & 87 & 9 & -90 & 25 & 80 & -57 \\ 50 & -89 & 18 & 75 & -75 & -18 & 89 & -50 & -50 & 89 & -18 & -75 & 75 & 18 & -89 & 50 \\ 43 & -90 & 57 & 25 & -87 & 70 & 9 & -80 & 80 & -9 & -70 & 87 & -25 & -57 & 90 & -43 \\ 36 & -83 & 83 & -36 & -36 & 83 & -83 & 36 & 36 & -83 & 83 & -36 & -36 & 83 & -83 & 36 \\ 25 & -70 & 90 & -80 & 43 & 9 & -57 & 87 & -87 & 57 & -9 & -43 & 80 & -90 & 70 & -25 \\ 18 & -50 & 75 & -89 & 89 & -75 & 50 & -18 & -18 & 50 & -75 & 89 & -89 & 75 & -50 & 18 \\ 9 & -25 & 43 & -57 & 70 & -80 & 87 & -90 & 90 & -87 & 80 & -70 & 57 & -43 & 25 & -9 \end{bmatrix} \tag{2.19}$$

$$H = \begin{bmatrix} 29 & 55 & 74 & 84 \\ 74 & 74 & 0 & -74 \\ 84 & -29 & -74 & 55 \\ 55 & -84 & 74 & -29 \end{bmatrix} \tag{2.20}$$

The quantisation procedure used in HEVC is essentially the same as that used in H.264/AVC, basically consisting of a non-linear discrete mapping of the coefficient values into integer quantisation indices. It is implemented as a multiplication by a constant, an addition of a rounding factor and a right shift controlled by a quantisation parameter (QP), which varies from 0 to 51. All transformed coefficients of a TB are quantised and inverse quantised depending on the QP value. Table 2.3 presents the constant $Q_{\mathrm{QP\%6}}$ values used in the quantisation calculation, where % is an operator that computes the remainder of the division between QP and 6. The quantisation is done according to Eq. (2.21), where L is the quantised coefficient (quantisation output), C is the transformed coefficient (quantisation input), *offset* is a rounding factor and N is the TB dimension [17]:

$$L = \frac{C \cdot Q_{\mathrm{QP\%6}} + \mathrm{offset}}{2^{21 + \frac{\mathrm{QP}}{6} - \log_2 N}} \tag{2.21}$$

Table 2.3 $Q_{\mathrm{QP\%6}}$ values used in the coefficient quantisation

	QP%6					
	0	1	2	3	4	5
$Q_{\mathrm{QP\%6}}$	26,214	23,302	20,560	18,396	16,384	14,564

Rate-distortion optimised quantisation (RDOQ) [18] has been adopted in HEVC as a non-normative quantisation optimisation technique. When the technique is enabled, the encoder computes four quantised coefficient candidates for each coefficient in each TB and then selects the best candidate in terms of R-D efficiency through repetitive processing.

After the transform and quantisation are finished, the coded block flag (CBF) indicates whether or not the TB includes residual information. If the TB includes residue information, the CBF value is set to 1; otherwise, it is set to 0.

2.6.2.4 Inverse Quantisation and Inverse Transform

The inverse quantisation (IQ) and inverse transform (IT) modules are responsible for performing the inverse operations of the quantisation and transform modules, as their names suggest.

The IQ is defined in Eq. (2.22), where CQ is the inverse quantised coefficient (IQ output), L is the quantised coefficient (IQ input) and N is the TB dimension [17]. The possible $\mathrm{IQ}_{\mathrm{QP}\%6}$ values are presented in Table 2.4, where % is an operator that computes the division remainder.

$$\mathrm{CQ} = \frac{L \cdot \mathrm{IQ}_{\mathrm{QP}\%6} \cdot 2^{\frac{\mathrm{QP}}{6}}}{2^{\log_2 N - 1}} \tag{2.22}$$

Once the inverse quantised coefficients are calculated, the IT corresponding to the TB dimension is applied to them, resulting in the inverse transformed residue prediction. The inverse transformation matrices are defined as the transposes of their corresponding forward transformation matrices. For example, the inverse transformation matrix for 16×16 TBs is the transpose of the matrix presented in Eq. (2.19). Similarly, the inverse transformation matrix of 4×4 DST is the transpose of the matrix in Eq. (2.20).

2.6.2.5 Entropy Coding

The entropy coding process of HEVC uses only one tool, which is also available in H.264/AVC: the context-adaptive binary arithmetic coding (CABAC). The CABAC receives the quantised coefficients, reorganises them according to one of the three available scanning orders (diagonal upright, horizontal and vertical), selects a

Table 2.4 $\mathrm{IQ}_{\mathrm{QP}\%6}$ values used in the coefficient quantisation

	QP%6					
	0	1	2	3	4	5
$\mathrm{IQ}_{\mathrm{QP}\%6}$	40	45	51	57	64	72

probability model for each syntactic element according to its context, updates the probability models and finally encodes the element. As the CABAC operation is not directly affected by any of the complexity reduction and scaling methods proposed in this book, it will not be described in further detail. More information about this module can be found in [9, 19].

2.6.2.6 In-Loop Filters

Two filtering steps are applied to the reconstructed samples (i.e. after adding the predicted samples to the reconstructed prediction residue), before writing them into the DPB, as shown in Fig. 2.4. The first filter is the deblocking filter (DBF), which was already present in the H.264/AVC standard, and the second filter is the sample adaptive offset (SAO), which is a new tool introduced in HEVC.

The DBF is applied to boundaries of CBs, PBs and TBs larger than 4×4 pixels [20]. Vertical and horizontal edges are filtered according to a border filtering strength, which varies from 0 (no filtering) to 4 (maximum filtering strength) and also depends on the border characteristics. The SAO, on the other hand, is applied to all samples of the image, classifying the reconstructed pixels into different categories and then reducing distortion by adding an offset to the pixels in each category [21]. This classification is performed taking into consideration the pixel intensity and edge properties, e.g. based on gradient.

During the HEVC standardisation process, a third filter called adaptive loop filter (ALF) was proposed. However, the ALF was not included in the final standard. Still, as part of the research work presented in this book was performed during the standardisation process, the ALF was still considered in some studies presented later in this book.

Since the in-loop filter operation is not directly affected by any of the methods proposed in this book, the DBF and SAO operations will not be further detailed in this chapter, but additional information can be found in [20, 21].

2.6.3 Test Conditions, Temporal Configurations and Profiles

As in the case of previous standards, the HEVC specification [22] describes the standard syntax and decoding procedures. The encoder can be freely implemented as long as the generated bitstream respects the syntax and decoding rules defined by the standard. In order to guide tests and proposals submitted during the standardisation process, the JCT-VC group coordinated the development of a reference software encoder and decoder, colloquially known as HM, and published the *common test conditions* (CTC) document [23], which defines a set of four temporal configurations to be used with the HM reference software for tests and comparisons of competing proposals for modification of the standard. Additionally, the standard also specifies three profiles, which are conformance points that define a set of tools or algorithms that can be implemented in the encoder according to its application (e.g. different profiles can be used for different types of device).

Four temporal configurations which differ in terms of temporal prediction from one another are defined in the CTC: *All Intra*, *low delay*, *low delay P* and *random access*. Furthermore, two profiles are defined for video coding: *Main* and *Main 10*. The combination of the four temporal configurations and the two profiles constitute the eight testing conditions used in the latest versions of HM [23]. The CTC document is presented in the Appendix B of this book.

2.6.3.1 Temporal Configurations

The encoding temporal configuration defines which frames can be used as references in the prediction process.

In the *All Intra* (AI) configuration, all images in the video sequence are encoded as instantaneous decoding refresh (IDR) pictures, which are pictures that contain only I slices and therefore may be the first picture in decoding order, since they make no reference to others. When encoding video according to this configuration, inter-frame prediction is not performed and there are no pictures stored in the reference lists. The QP value is held unchanged during the encoding of the entire video sequence. Figure 2.16 shows an example of a video sequence encoded with the AI configuration, where the encoding order is presented at the top of each picture. In this configuration, the encoding and the display order are the same.

In the *low delay* (LD) configuration, only the first image in the GOP is encoded as an IDR picture. The remaining images are encoded as generalised P and B pictures (GPB), which are B pictures that only allow using reference frames that appear before the current frame in display order. In other words, a GPB picture is able to use only reference pictures whose encoding order is smaller than the current picture. Both reference lists (*List 0* and *List 1*) are identical. Figure 2.17 shows an example of a GOP encoded with the LD configuration. The number associated with each picture represents the encoding order. Notice that even though bidirectional prediction is allowed, each picture only uses previous frames (in display order) as references. The *low delay P* (LDP) configuration is a variation of the LD configuration that functions in a very similar way. The only difference is that only unidirectional prediction is allowed. This way, only one reference list (*List 0*) is maintained in the DPB.

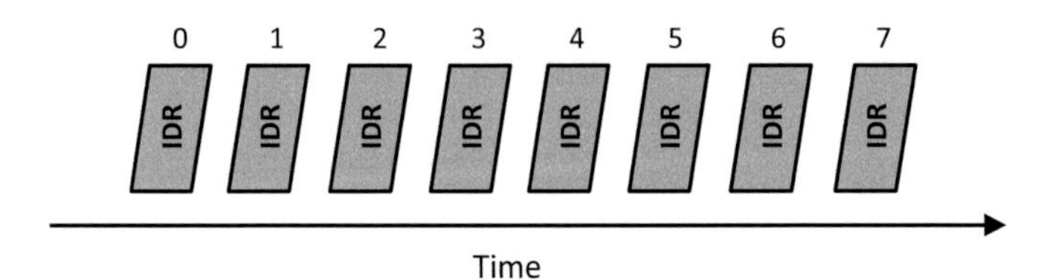

Fig. 2.16 Graphical presentation of the AI configuration

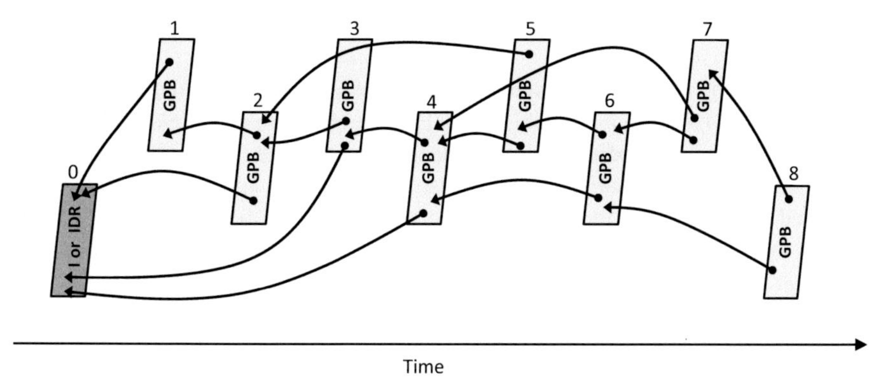

Fig. 2.17 Graphical presentation of the LD configuration

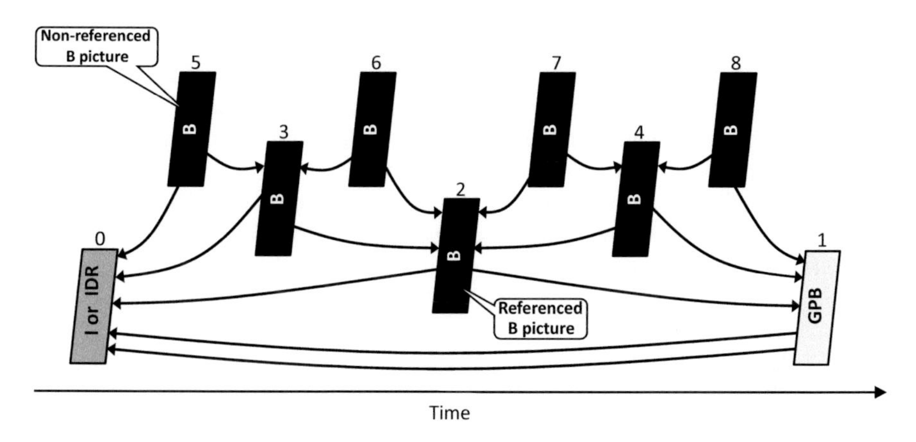

Fig. 2.18 Graphical presentation of the RA configuration

Finally, the *random access* (RA) configuration is characterised by the utilisation of a temporal hierarchical B structure, which is illustrated in Fig. 2.18. The number associated to each picture represents the encoding order and the picture display order is represented from left to right. The first picture in a video sequence is encoded as an IDR, and the remaining intra pictures are encoded as non-IDR intra pictures, characterising an open GOP. An open GOP means that frames outside the current GOP can be used as references. Every frame between two intra-frames is encoded as a B picture. In the first temporal layer of Fig. 2.18, a GPB picture is used. In the second and third temporal layers, ordinary B pictures are used (i.e. bidirectional inter-frame prediction using both reference lists, which are not necessarily identical). Finally, the fourth and last temporal layer also uses ordinary B pictures, but these cannot reference each other.

2.6.3.2 Profiles

At the beginning of the HEVC standardisation, two profiles were defined: the *low complexity* (LC) and the *high efficiency* (HE). The two profiles were different from each other in terms of coding tools and enabled functionalities, but several changes happened until the finalisation of the standard, so that at the end only one difference remained between them: the internal bit depth. For this reason, the LC and the HE profiles were removed from the standard, and two new profiles were defined: *Main* and *Main 10*. The *Main* profile uses all tools described in the HEVC specification draft [22], most of which were covered in this chapter. The *Main 10* profile contains the same tools of *Main*, but the bit depth for both luminance and chrominance samples is set to 10 bits instead of 8 bits.

When the *Main* profile is used with a 10-bit source, each source sample is scaled to an 8-bit value prior to encoding in a process called internal bit depth decrease (IBDD). The scaling is obtained by applying the function $y=(x+2)/4$ to the input value x and clipping the result y to the [0, 255] range. Oppositely, when the *Main 10* configuration is used with an 8-bit source, each sample is scaled to a 10-bit value before encoding by applying the $y=4*x$ function to the input value x. This process is called internal bit depth increase (IBDI), and it allows greater precision in the video codec operations (i.e. improved encoding efficiency) at the cost of an increase in memory requirements, mainly to store reference pictures in the DPB.

All experiments presented in this book were performed using the *Main* profile (or the *low-complexity* profile, in early versions of HM).

2.7 Computational Complexity of HEVC

As part of the research presented in this book, the computational complexity of the HM encoder was extensively characterised and quantified through several tests and comparisons, as presented in [24]. Such experiments concluded that the HM encoder presents a computational complexity increase varying from 9 % up to 502 % in comparison to the H.264/AVC encoder (high profile), depending on the HEVC encoding configuration. Detailed experiments and results are presented in Chap. 4.

The computational complexity and implementation of both HM encoder and decoder were also analysed in [25]. It was observed that in the AI encoder configuration, the most time-consuming modules are the transform and quantisation (24.4 %), due to the RDOQ technique, the intra-prediction (16.6 %) and the entropy coding (2.2 %). In the RA encoder configuration, the ME takes up a significant portion of encoding time (38.8 % for SAD calculations and 19.8 % for fractional pixel search refinement). The decoding computational complexity is dominated by the IT (15.9 %) and the filters (12.9 %) in the AI configuration and by the MC (24.8 %) and the filters (12.4 %) in the RA configuration.

Besides the encoding modules of HEVC, the encoding structures presented in this chapter, especially the frame partitioning structures, are also responsible for the high computational complexity involved in the encoding task. With a naïve implementation of the RDO process without simplifications, the process of defining the optimal combination of CBs, PBs and TBs would involve encoding each CTB using all possibilities allowed by the encoder, comparing their R-D costs and finally choosing the best one, which characterises a very complex procedure.

Also as a part of the research presented in this book, an analysis of the computational complexity for defining the frame partitioning structures of HEVC was performed. This analysis allowed concluding that the nature of such structures leads to nested encoding loops, such that CBs at large tree depths are encoded inside CBs at smaller tree depths, significantly increasing the encoding computational complexity. Detailed experiments and results of this analysis are also presented in Chap. 4.

References

1. I.E. Richardson, *Video Codec Design: Developing Image and Video Compression Systems* (Wiley, New York, 2002)
2. I.E. Richardson, *H.264/AVC and MPEG-4 Video Compression – Video Coding for Next-Generation Multimedia* (John Wiley and Sons, Chichester, 2003)
3. M. Ghanbari, *Standard Codecs: Image Compression to Advanced Video Coding* (The Institute of electrical Engineers, London, 2003)
4. Y. Shi, H. Sun, *Image and Video Compression for Multimedia Engineering: Fundamentals, Algorithms and Standards* (CRC Press, Boca Raton, 1999)
5. G. Bjontegaard, *Calculation of Average PSNR Differences Between RD-Curves* (Texas, Austin, 2001)
6. G.J. Sullivan, T. Wiegand, Rate-distortion optimization for video compression. IEEE Signal Process. Mag. **15**, 74–90 (1998)
7. T. Wiegand, G.J. Sullivan, G. Bjontegaard, A. Luthra, Overview of the H.264/AVC video coding standard. IEEE Trans. Circ. Syst. Video Technol. **13**, 560–576 (2003)
8. A. Ortega, K. Ramchandran, Rate-distortion methods for image and video compression. IEEE Signal Process. Mag. **15**, 23–50 (1998)
9. G.J. Sullivan, J. Ohm, H. Woo-Jin, T. Wiegand, Overview of the high efficiency video coding (HEVC) standard. IEEE Trans. Circ. Syst. Video Technol. **22**, 1649–1668 (2012)
10. ISO/IEC-JCT1/SC29/WG11, High Efficiency Video Coding (HEVC) Test Model 13 (HM 13) Encoder Description, Geneva, Switzerland (2013).
11. J. Lainema, F. Bossen, H. Woo-Jin, M. Junghye, K. Ugur, Intra coding of the HEVC standard. IEEE Trans. Circ. Syst. Video Technol. **22**, 1792–1801 (2012)
12. ISO/IEC-JCT1/SC29/WG11, Encoder improvement of unified intra prediction, Guangzhou, China (2010).
13. X. Xu, Y. He, Comments on Motion Estimation Algorithms in Current JM Software, doc. JVT-Q089 (2005).
14. N. Purnachand, L.N. Alves, A. Navarro, Improvements to TZ search motion estimation algorithm for multiview video coding, in *2012 19th International Conference on Systems, Signals and Image Processing (IWSSIP)* (2012), pp. 388–391.
15. P. Helle, S. Oudin, B. Bross, D. Marpe, M.O. Bici, K. Ugur, J. Jung, G. Clare, T. Wiegand, Block merging for Quadtree-based partitioning in HEVC. IEEE Trans. Circ. Syst. Video Technol. **22**, 1720–1731 (2012)
16. ISO/IEC-JTC1/SC29/WG11, Samsung's response to the call for proposals on Video compression technology, Dresden, Germany (2010).

17. G. RyeongHee, L. Yung-Lyul, N-level quantization in HEVC, in *2012 IEEE International Symposium on Broadband Multimedia Systems and Broadcasting (BMSB)* (2012), pp. 1–5.
18. Y. Y. M. Karczewicz, I. Chong, Rate distortion optimized quantization, in *VCEG-AH21*, Antalya, Turkey (2008).
19. D. Marpe, H. Schwarz, T. Wiegand, Context-based adaptive binary arithmetic coding in the H.264/AVC video compression standard. IEEE Trans. Circ. Syst. Video Technol. **13**, 620–636 (2003)
20. A. Norkin, G. Bjontegaard, A. Fuldseth, M. Narroschke, M. Ikeda, K. Andersson, Z. Minhua, G. Van der Auwera, HEVC deblocking filter. IEEE Trans. Circ. Syst. Video Technol. **22**, 1746–1754 (2012)
21. F. Chih-Ming, E. Alshina, A. Alshin, H. Yu-Wen, C. Ching-Yeh, T. Chia-Yang, H. Chih-Wei, L. Shaw-Min, P. Jeong-Hoon, H. Woo-Jin, Sample adaptive offset in the HEVC standard. IEEE Transactions on Circuits and Systems for Video Technology **22**, 1755–1764 (2012)
22. ISO/IEC-JCT1/SC29/WG11, High Efficiency Video Coding (HEVC) text specification draft 10, Geneva, Switzerland (2013).
23. ISO/IEC-JCT1/SC29/WG11, Common test conditions and software reference configurations, Geneva, Switzerland (2012).
24. G. Correa, P. Assuncao, L. Agostini, L.A. da Silva Cruz, Performance and computational complexity assessment of high efficiency video encoders. IEEE Transactions on Circuits and Systems for Video Technology **22**, 1899–1909 (2012)
25. F. Bossen, B. Bross, K. Suhring, D. Flynn, HEVC complexity and implementation analysis. IEEE Transactions on Circuits and Systems for Video Technology **22**, 1685–1696 (2012)

Chapter 3
State of the Art on Computational Complexity of Video Encoders

This chapter presents an overview of the state-of-the-art research on computational complexity of video encoders. First, a description of current methods for modelling the computational complexity of state-of-the-art video encoders is presented. Then, methods for reducing and scaling the expenditure of computational resources in video codecs are presented. Finally, the conclusions of the chapter highlight the research challenges and open topics not fully addressed in the available literature.

3.1 Modelling Computational Complexity

Estimating if the encoding operations will exceed the system's available computational resources is an important problem in the design of complexity-aware video coding applications. As the computational resources cannot be retrieved once spent, the system must be able to detect such cases without the need of performing the actual operations. Therefore, modelling computational complexity of video codecs is necessary to allow estimating the computational complexity required by a certain operation prior to its execution. Nevertheless, this is not a trivial task because video codecs are composed of several interdependent tools with parameterised functionalities that can influence significantly the level of computational complexity. Furthermore, according to [1], in the case of a complex multimedia system, an overall view may not be enough to understand interrelated modules as a whole; even though detailed information about all the constitutive functional modules and their theoretical complexity may be available, the overall codec behaviour and complexity are very difficult to be understood, because they depend on the input data characteristics.

In the recent past, modelling the computational complexity of video encoders and decoders was investigated in several works, some of which are presented in the next subsections. Even though there are still few works published on the topic

G. Corrêa et al., *Complexity-Aware High Efficiency Video Coding*,
DOI 10.1007/978-3-319-25778-5_3

aiming at the HEVC encoding/decoding process (Sect. 3.1.2), many works focusing on H.264/AVC are available in the technical literature (Sect. 3.1.1).

3.1.1 Computational Complexity Modelling for H.264/AVC

In [2–4], the authors focus on modelling the computational complexity of inter-frame prediction operations of the H.264/AVC decoder due to their large share of the overall complexity. Since different inter-prediction modes demand different computational complexities, this was taken into account to estimate the overall decoding complexity in [2]. The number of MVs was used in [3] to estimate the decoding complexity of a macroblock (MB). According to the experiments described in [3], the decoding computational complexity is directly proportional to the number of MVs used in a MB. In [4], the number of interpolation filter operations was used for the complexity estimation. According to the authors, a MB with fewer interpolation filter operations has lower decoding complexity.

The computational complexity of the inter-frame prediction is modelled in [5] for the H.264/AVC decoder. The motion compensation (MC) complexity is modelled for each inter-prediction mode as a linear function of the number of cache misses, the number of interpolations and the number of MVs per MB. Then, an H.264/AVC encoder equipped with the developed model is proposed, aiming at estimating the decoding complexity and choosing the mode that best meets a target decoding complexity.

There are also some approaches that deal with the computational complexity of video encoders and decoders based on their basic data structures. In H.264/AVC, these structures are blocks, MBs, frames and GOPs. In [6], the computational complexity of the H.264/AVC encoding process for a frame is estimated based on the computational complexity of encoding the previous frame. A frame is divided into groups of MBs, which are further used by the complexity scaling algorithm, which assigns different complexity budgets to each group. The complexity of the frame is computed by simply adding the complexity of each group of MBs, which in turn is computed by summing up the complexity of each MB.

Some works have also proposed computational complexity models for video encoders that are dependent upon configuration parameters used for complexity scaling. The H.264/AVC encoder complexity was modelled in [7] by first dividing it into three major parts: ME (motion estimation), PRECODING (transforms, quantisation and picture reconstruction) and EC (entropy coding). The ME computational complexity is given by Eq. (3.1), where C_{SAD} is the computational complexity of a SAD operation and λ_{ME} is the number of SAD calculations per frame, which is the complexity scaling parameter for the ME module. The PRECODING complexity is given by Eq. (3.2), where C_{NZMB} is the PRECODING computational complexity of one nonzero MB and λ_{PRE} is the number of nonzero MBs in a frame, which is the complexity scaling parameter for the PRECODING module. Here, a NZMB is a MB with nonzero DCT coefficients after the quantisation. Finally, the

EC computational complexity is given by Eq. (3.3), and it is proportional to the bit rate R. In Eq. (3.3), C_{BIT} is the per-bit complexity associated to EC and S is the size of the picture, which is needed because R represents the coding bit rate in bits per pixel. The total computational complexity C of the video encoder is given by Eq. (3.4), where λ_{F} is the video frame rate. The model represents a complexity-scalable architecture for video encoding, for which the computational complexity is scaled by the parameters λ_{ME}, λ_{PRE} and λ_{F}:

$$C_{\mathrm{ME}} = \lambda_{\mathrm{ME}} \cdot C_{\mathrm{SAD}} \tag{3.1}$$

$$C_{\mathrm{PRE}} = \lambda_{\mathrm{PRE}} \cdot C_{\mathrm{NZMB}} \tag{3.2}$$

$$C_{\mathrm{ENC}} = S \cdot R \cdot C_{\mathrm{BIT}} \tag{3.3}$$

$$C\left(R;,,\lambda_{\mathrm{ME}};,,\lambda_{\mathrm{PRE}};,,\lambda_{\mathrm{F}}\right) = \lambda_{\mathrm{F}} \cdot \left(\lambda_{\mathrm{ME}} \cdot C_{\mathrm{SAD}} + \lambda_{\mathrm{PRE}} \cdot C_{\mathrm{NZMB}} + S \cdot R \cdot C_{\mathrm{BIT}}\right) \tag{3.4}$$

Similarly, in [8] the computational complexity of the H.264/AVC encoder is also modelled by dividing it into three modules: ME, PRECODING and EC. Two models are proposed: the first one considers that the RDO decision is enabled, while the second considers that the RDO process is off. When RDO is used, the encoder executes the ME, transform, quantisation and entropy coding for each mode tested before choosing the best option. This is represented by Eq. (3.5), where C_{RDOon} is the encoding complexity of the frame; $C^i_{16\times16}$–$C^i_{\mathrm{I}4\times4}$ are the complexities associated to each prediction mode at the ith MB; C^i_M is the complexity of the transform, quantisation and entropy coding for each mode at the ith MB; and C^i_{PRE} and C^i_{EC} are the complexities of the PRECODING and EC modules after the decision mode is performed at the ith MB. When RDO is off, the mode decision is based on the results of the ME process only, so that the model in Eq. (3.6) does not include the C^i_M element. In Eq. (3.6), C_{RDOoff} represents the total encoding complexity of the frame when RDO is off:

$$C_{\mathrm{RDOon}} = \sum_{i=0}^{M-1}\left(C^i_{16\times16} + C^i_{16\times8} + C^i_{8\times8} + C^i_{I16\times16} + C^i_{I4\times4} + C^i_{M}\right) + \sum_{i=0}^{M-1}\left(C^i_{\mathrm{PRE}} + C^i_{\mathrm{ENC}}\right) \tag{3.5}$$

$$C_{\mathrm{RDOoff}} = \sum_{i=0}^{M-1}\left(C^i_{16\times16} + C^i_{16\times8} + C^i_{8\times16} + C^i_{8\times8} + C^i_{I16\times16} + C^i_{I4\times4}\right) + \sum_{i=0}^{M-1}\left(C^i_{\mathrm{PRE}} + C^i_{\mathrm{ENC}}\right) \tag{3.6}$$

In [9], the computational complexity of H.264/AVC encoders is modelled based on the complexity of each prediction mode (*SKIP*, *Inter 16×16*, *Inter 16×8*, *Inter 8×16*, *Inter 8×8*, *Inter 8×4*, *Inter 4×8*, *Inter 4×4*, *P8×8*, *I4×4*, *I16×16*). The model for the full computational complexity of a P frame is calculated as in Eq. (3.7), where W^{RD}_i indicates the normalised R-D computational cost for mode i, W^{ME}_i denotes the factor of single-reference ME for the inter modes, N_{MB} is the number of MBs in the frame, N_R is the number of reference frames and W^O is the complexity of the remaining encoder modules. The computational complexity of the *Inter 16×16* mode with a single-reference frame was used as basis for the normalisations:

$$C^{\text{Full}}_{\text{Frame}} = N_{\text{MB}} \cdot \begin{bmatrix} W_0^{\text{RD}} + \sum_{i=1}^{7}\left(W_i^{\text{RD}} + W_i^{\text{ME}} \cdot N_{\text{R}}\right) + \\ W_8^{\text{RD}} + W_9^{\text{RD}} + W_{10}^{\text{RD}} + W^{\text{O}} \end{bmatrix} \quad (3.7)$$

The H.264/AVC decoding computational complexity is modelled in [10] using an approach where each decoding module (DM) is separately modelled. Entropy decoding complexity is modelled as the product of bit decoding complexity and the number of bits involved, as shown in Eq. (3.8), where C_{vld} is the DM complexity, k_{bit} is the average number of cycles for decoding one bit and n_{bit} is the number of bits for a frame. Side information preparation, including MB sum/clip and DBF strength calculation, is modelled as in Eq. (3.9), where C_{sip} is the DM complexity, k_{MBsip} represents the average clock cycles for side information preparation per MB and n_{MB} is the number of MBs in a frame. Inverse transform complexity is modelled in Eq. (3.10), where $k_{MBitrans}$ is a constant that describes the complexity of MB dequantisation and inverse transform and n_{nzMB} is the number of nonzero MBs per picture. Intra-frame prediction is modelled in Eq. (3.11), as the product between the average complexity of intra-frame prediction in one intra-coded MB ($k_{intraMB}$) and $n_{IntraMB}$ is the number of intra MBs per frame. The MC complexity model is given in Eq. (3.12), where k_{half} is the average complexity required to conduct a 6-tap Wiener filtering and n_{half} is the number of filterings to decode a frame. The DBF is modelled as in Eq. (3.13), where k'_α and k'_β are the complexities to perform strong and normal filterings, respectively, and n_α and n_β are the numbers of strong and normal filterings, respectively. Finally, the total complexity required to decode a frame is expressed as in Eq. (3.14), where $k_{CU(DM)}$ is the complexity to decode a basic coding unit (e.g. a MB) in a particular module and $N_{CU(DM)}$ is the number of basic coding units to be decoded by that module:

$$C_{\text{vld}} = k_{\text{bit}} \cdot n_{\text{bit}} \quad (3.8)$$

$$C_{\text{sip}} = k_{\text{MBsip}} \cdot n_{\text{MB}} \quad (3.9)$$

$$C_{\text{itrans}} = k_{\text{MBitrans}} \cdot n_{\text{nzMB}} \quad (3.10)$$

$$C_{\text{Intra}} = k_{\text{IntraMB}} \cdot n_{\text{IntraMB}} \quad (3.11)$$

$$C_{\text{mcp}} = k_{\text{half}} \cdot n_{\text{half}} \quad (3.12)$$

$$C_{\text{dblk}} = k'_{\alpha} n_{\alpha} + k'_{\beta} n_{\beta} \tag{3.13}$$

$$C_{\text{frame}} = \sum_{\text{DM}} C_{\text{DM}} = \sum_{\text{DM}} C_{\text{CU(DM)}} \cdot N_{\text{CU(DM)}} \tag{3.14}$$

3.1.2 Computational Complexity Modelling for HEVC

As HEVC is still a recent standard, only a few works focusing on modelling its computational complexity are currently available in the technical literature [11, 12].

In [11], an analysis based on data mining is used to construct a computational complexity model for HEVC using linear regression. The model uses encoder parameters to determine the time savings provided when using different configurations in comparison to a baseline case. The model is shown in Eq. (3.15), where *G*, *H*, *I*, *J*, *K*, *L* and *M* are model parameters, *height* is the number of pixels lines per frame, *BR* is the target bit rate, *LCU* is the dimension of the CTU, *CUD* is the maximum coding tree depth allowed, *TUD* is the maximum RQT depth allowed in inter-predicted CUs and *TS* is the time saving obtained when using the current encoding parameters, in comparison to a reference configuration:

$$\text{TS} = \frac{H \cdot \text{height} + I \cdot \text{BR} - J \cdot \text{LCU} - K \cdot \text{CUD} - L \cdot \text{TUD} + M}{G} \tag{3.15}$$

In [12], the authors propose a load-balancing algorithm for parallel scheduling in HEVC. A model is used to predict the complexity of each slice and tile, in order to allocate the same amount of work for each core before the encoding process starts. The computational complexity of each slice or tile is computed by summing up the complexities of all CTUs, which are individually computed as in Eq. (3.16), where *CC* is the computational complexity of encoding a CTU, *CEM*(*s*,*m*) is the normalised complexity of each prediction mode *m* in a CU of size *s* (measured through offline encodings) and *CHK*(*s*,*m*) is the information of whether or not the mode is tested for the CTU. The prediction mode *m* and the CU size *s* belong to sets *S* and *M*, given by Eq. (3.17) and Eq. (3.18), respectively. If *m* is a valid mode for the current CU size *s*, *CHK*(*s*,*m*) is set to 1; otherwise, it is set to 0, as shown in Eq. (3.19):

$$\text{CC} = \sum_{s \in S} \sum_{m \in M} \text{CEM}(s,m) \cdot \text{CHK}(s,m) \tag{3.16}$$

$$S = \{64 \times 64, 32 \times 32, 16 \times 16, 8 \times 8\} \tag{3.17}$$

$$M = \{\text{MERGE}, \text{INTER}, \text{INTRA}\} \tag{3.18}$$

$$\mathrm{CHK}(s,m) = \begin{cases} 1, & \text{if valid}(s,m) \\ 0 & \text{otherwise} \end{cases} \tag{3.19}$$

3.2 Computational Complexity Reduction

Low-complexity algorithms for video coding and decoding have been proposed in the last years as an attempt to enable the use of compressed video in complexity-constrained platforms. As the encoder's computational complexity is much higher than the decoder's, most works focus on its modules and features.

3.2.1 Computational Complexity Reduction for H.264/AVC

Most works found in the literature focus on the ME and MD tasks because they are the most complex operations in the encoding process and so provide ampler room for decreasing the encoding computational complexity. As ME and MD are operations common to most video coding standards, most of the techniques presented in the next subsections can be applied not only to H.264/AVC but also to HEVC with some adaptations.

3.2.1.1 Motion Estimation

Several algorithms have been proposed for searching candidate blocks, and the algorithmically simplest but most demanding in terms of computation needs is the *full search* (FS). In FS, the best match is found by searching all possible candidate blocks in a search area (SA) of the reference frame. However, the computational complexity involved in this process is very high and usually faster approaches are used.

Several *fast motion estimation* (FME) techniques, which have been widely applied to H.264/AVC but were actually designed for use in any video coding standard that includes ME, have been proposed in the literature in the last decade. These techniques are classified into two categories here: (a) those which decrease the number of candidate blocks in the SA and (b) those which decrease the computational complexity required to compare such blocks to the current block (i.e. distortion measure computation, such as SAD or MSE).

Examples of the first category are the *Three-Step Search* (TSS) [13], the *Block-Based Gradient Descent Search* [14], the *New Three-Step Search* [15], the *Diamond Search* [16, 17], the *Hexagon-Based Search* [18], the *One at a Time Search* [19] and the *Dual Cross Search* [20]. All of them are suboptimal algorithms with R-D performance equal to or smaller than that obtained from FS, with the advantage of significantly decreasing the ME computational complexity.

Methods from the second category include techniques such as subsampling [21], *Partial Distortion Search* [22], *Normalised Partial Distortion Search* [23, 24] and the *Successive Elimination Algorithm* [25]. Even though these methods are capable of maintaining good R-D performance, their complexity reduction factor is limited.

There are still other works that reduce the ME computational complexity by applying different techniques. In [26], the authors propose a strategy based on sorting the MVs and coding modes such that the decision process is stopped when the rate required to encode a MV or coding mode exceeds the average of any previous MV or mode tested, thus skipping the evaluation of some cases. In [27], a controller is added to the encoder to extract signal statistics from the motion search and use them to dynamically configure the ME parameters, such as the number of reference frames, tested block modes and SA. A method for estimating which MBs can be encoded in *SKIP* mode without trying the other modes is proposed in [28]. The authors propose a Bayesian framework using the R-D cost difference between coding and skipping a MB as the single decision feature. Savings in processing time of more than 80 % for low-motion sequences are claimed by the authors at a cost of small quality decrease.

3.2.1.2 Mode Decision

The MD process involves different operations from various encoder modules. If RDO is enabled, then every possible coding mode is tried and the R-D performance results obtained using all the modes are compared in order to find and select the best mode. Many works based on heuristics have been proposed to decrease the amount of modes to be tested during both intra-frame and inter-frame prediction.

In [29–32], the authors propose fast algorithms to select a subset of intra-frame prediction modes to be compared in the MD stage. The authors in [33] and [34] claim that not all intra-modes need to have the R-D performance evaluated, especially those with a high *Sum of Absolute Transformed Differences* (SATD) value, since they tend to result in high encoding bit rates. Therefore, only those modes with an SATD value smaller than a threshold are evaluated, and the remaining modes are ignored, reducing the volume of computation to be performed.

In [35], the transforms are applied to the intra-frame prediction residue, and the transformed coefficients are analysed in order to detect which is the best mode to be used. In [36] the low-frequency transformed coefficients for each MB are analysed in order to decide the best intra-frame block size (4×4 or 16×16).

A two-step algorithm for intra-frame MD was proposed in [37] based on heuristics. The first step consists in performing intra-frame prediction for every mode in both 16×16 and 4×4 prediction block sizes and computing the distortion between the original and predicted blocks. The mode that results in the smallest distortion is used as the best mode for its corresponding block size. In the second step, the MB homogeneity is computed to decide whether a 16×16 or a 4×4 block size should be used. A pruned DCT is applied to the original MB in order to compute a subset of coefficients, which are used to quantify the homogeneity level of the MB, which will then guide the block size choice. Homogeneous areas of the image must be encoded with large blocks, while heterogeneous areas must be encoded using small blocks.

In [38], a hierarchical fast inter-frame MD based on the evaluation of temporal stationarity, texture homogeneity and block border strength was proposed. The method is divided into three steps. In the first one, stationarity is detected by calculating the distortion between a MB and its co-localised MB in the reference frame. If stationarity is detected, the MB is encoded as *SKIP* and the MD process is terminated. Otherwise, the second step takes place and the homogeneity of the MB is computed as in [37] in order to decide if the ME is performed with large or small blocks. In the third and final step, the intensity variation of luminance samples along the edges of possible blocks is calculated in order to detect the better block shape for inter-frame prediction, eliminating the need of testing all the remaining block shapes. This method was integrated with the fast intra-frame decision proposed in [37], building a complete MD scheme for H.264/AVC [39].

According to [40], only 3 % of the modes chosen in P frames are intra-prediction modes, on average. This way, the evaluation of inter-frame modes prior to the evaluation of intra-frame modes is proposed in [40] as a way of reducing the necessary computational resources without significant decreasing image quality or bit rate. In [41] a set of heuristics is presented for speeding up the complete MD process (i.e. both intra and inter-frame predictions). The authors observe that homogeneous and static regions are mostly encoded with inter 16×16 and *SKIP* modes, so that a Sobel operator is used to detect homogeneity and the SAD calculation is used to detect stationarity. The proposed algorithm reduces the encoding complexity through an RDO early termination when one of the specific conditions (homogeneity or stationarity) is true. If neither is true, all the remaining prediction modes are evaluated.

In [42] the inter-frame MD complexity is decreased by analysing the spatial continuity of the motion field, which is generated by ME using 4×4 pixel blocks. A set of experiments have shown that video regions with high motion continuity present a higher probability of being encoded with inter 16×16, 16×8 and 8×16 modes, while regions with low-motion continuity are mostly encoded with inter 8×8 and smaller block sizes. As in [41], the Sobel operator is also used in [42] to identify the borders of objects in an image.

Table 3.1 summarises the existing computational resource-saving strategies for H.264/AVC. Categories "FME Cand.", "FME Dist." and "FME Other" correspond to approaches that (a) decrease the number of candidate blocks in the SA, (b) decrease the computational complexity of distortion calculation and (c) apply mixed techniques. Categories "Intra MD", "Inter MD" and "Full MD" correspond to FMD algorithms for intra-frame prediction, inter-frame prediction and both intra-/inter-frame predictions, respectively. Whenever available in the referenced paper, the amount of computational complexity reduction achieved with each approach is

Table 3.1 Computational complexity reduction strategies for H.264/AVC

Category	Approach	Reference	Complexity reduction
FME Cand.	Three-step search	[13]	–
	Block-based gradient descent search	[14]	–
	New three-step search	[15]	–
	Diamond search	[16, 17]	–
	Hexagon-based search	[18]	–
	One at a time search	[19]	–
	Dual cross search	[20]	–
	Subsampling	[21]	–
FME Dist.	Partial distortion search	[22]	–
	Normalised partial distortion search	[23, 24]	45–65 %[a]
	Successive elimination algorithm	[25]	–
FME other	Sorting MVs and modes	[26]	90 %
	Dynamic ME configuration	[27]	75 %[a]
	Bayesian early termination	[28]	80 %
Intra MD	Selective MD	[29]	82 %
	Selective MD	[30]	42–52 %
	Selective MD	[31]	–
	Selective MD	[32]	60 %
	Selective SATD-based MD	[33]	79 %
	Selective SATD-based MD	[34]	85 %[b]
	Transform coefficient analysis	[35]	20 %
	Transform coefficient analysis	[36]	52 %[b]
	Two-step hierarchical MD	[37]	99.3 %[c]
Inter MD	Three-step hierarchical MD	[38]	99.4 %[c]
	Motion field analysis	[42]	50 %
Full MD	Hierarchical MD	[39]	99.3 %[c]
	Inter MD before intra MD	[40]	30 %
	Homogeneity and stationarity detection	[41]	30 %

[a]Time reduction for the ME module only
[b]Time reduction for the intra-frame MD only
[c]Reduction in the overall number of tests performed under the RDO scheme

shown in the rightmost column of Table 3.1. These results are generally computed by comparing the algorithm proposed by the authors with the reference software or recommendation model and calculating the encoding time reduction. Exceptions are pointed out in the footnotes of Table 3.1.

It is important to observe that most of the approaches cannot be directly compared with one another in terms of complexity reduction because different encoder versions, encoding configurations, set-ups and experimental conditions (e.g. processor, memory, etc.) were used. Even so, the table provides useful information to the researcher interested in identifying the best complexity reduction strategies.

3.2.2 Computational Complexity Reduction for HEVC

Even though the HEVC standard has been recently launched, there are already several works published in the literature presenting methods to reduce the encoder's computational complexity. Most of these works aim at decreasing the computational complexity involved in the definition of the new frame partitioning structures, especially the coding trees, PUs and RQTs, and apply different techniques to determine the best configuration without testing all possibilities using an RDO process, as described in the next subsections.

3.2.2.1 Coding Tree Structure Determination

Fast algorithms for determining the coding tree structure are the most commonly found approaches to reduce the computational complexity of HEVC encoders, since the CU is the basic encoding unit and the coding tree determination is one of the most complex tasks of the encoding process, as Chap. 4 will show in detail.

A fast splitting and pruning method for intra-coding is proposed in [43]. It was found that the main contributor to the computational complexity of intra-prediction in HM is the calculation of R-D costs for deciding intra-prediction modes of CUs at all possible coding tree depths using RDO. In order to reduce these computational costs, the method uses a low-complexity R-D cost calculation to decide whether or not the CU splitting and pruning processes must be performed. The statistical parameters used in the Bayes-based choice of the best mode are periodically updated online so as to adapt to the changing characteristics of the video sequence. Experimental results have shown that the method is able to decrease the intra-coding computational complexity by 50 % with a BD-rate increase of 0.6 %. However, the method is only applicable to intra-predicted CUs, which are a minority in the configurations that use inter-frame prediction.

Two fast RDO techniques for HEVC are proposed in [44] aiming at saving computational resources at small R-D performance loss. The first method, the *Top Skip*, avoids checking the R-D cost for large blocks when they are unlikely to be chosen. A starting CTU depth is selected based on the minimum depth of the co-localised CTU in the reference frame. The second method, the *Early Termination*, avoids checking smaller blocks unlikely to be selected. The algorithm stops the CU splitting process if the best R-D cost is already lower than a given threshold. The threshold is adaptively computed based on the standard deviation of R-D costs relative to the CUs at spatially and temporally neighbouring CTBs. When both methods are integrated in the RDO process, a computational complexity reduction of 40 % is achieved for the whole encoding process with an average BD-rate increase of 1.9 %.

The methods proposed in [45] and [46] use information from intermediate encoding steps to determine if a CU is split into smaller CUs. In [45], a depth range selection mechanism is proposed. If the best prediction mode found for a determined CU is the *SKIP*, the splitting process is halted. In [46], the ratio between the

R-D costs in the current and upper-depth CUs is calculated and compared to thresholds in order to early terminate the CU splitting process. The methods [45] and [46] provide a computational complexity reduction of 48 % and 38 %, respectively, and average BD-rate increases of 1.7 % and 1.2 %, respectively.

A method based on motion divergence analysis is proposed in [47] to early terminate the splitting process of CUs. Before being encoded, each frame is downsampled and the optical flow of it is estimated based on the frame MVs in order to determine the motion divergence features. Then, for each CU, the motion divergence is evaluated as the variance of the optical flows of the current CU and its sub-CUs. The method yielded a computational complexity reduction of 43 % at the cost of a BD-rate increase of 1.9 %.

In [48], temporal correlation in neighbouring frames is exploited to reduce the number of quadtree splitting decisions. Based on the tree depth used in the co-localised CU in the previous frame and its neighbouring CUs, the encoder decides whether or not to split the current CU into sub-CUs. Additionally, an early *SKIP* mode decision at the prediction stage is performed to further reduce the computational complexity. Experimental results show that the method achieves a computational complexity reduction varying from 20 to 33 %, depending on the encoder configuration. BD-rate increases varied from 0.1 to 0.47 %.

A fast coding tree depth decision based on spatial and temporal correlation is proposed in [49]. The authors claim that since successive frames are strongly correlated, especially with the high frame rates lately used in video sequences, the final depth information or split structure of co-localised coding trees in neighbouring frames is also highly correlated. The algorithm first determines the depth search range according to the level of similarity between neighbouring CTBs. Three classes of similarity are defined (high, medium, low). Then, once the depth range is settled, the depth levels at which spatial partitioning will be tried can be derived by selecting depths with high probability of occurrence and excluding low probability depths. Experiments have shown that the method is able to reduce the encoding computational complexity in 25 % with an increase of 0.16 % in bit rate. No results in terms of BD measures were made available.

In [50], a fast coding tree depth decision algorithm which operates in both frame level and CU level is proposed for computational complexity reduction of HEVC encoding. The algorithm focuses only on the inter mode early determination, since, according to the authors, mismatches in intra CU sizes would result in too large PSNR drops or bit rate increases. At frame level, the main idea is to skip those depths which are rarely used in the reference frames. The number of CUs encoded at a certain tree depth in a frame is compared to a threshold in order to detect its level of usage. The CU part of the method relies on the fact that motion and texture detail of one particular part of the image tends to stay the same from one frame to another. By checking the spatial and temporal neighbouring CUs of a certain CU, the candidate CU depth can be predetermined. An average decrease of 45 % in the ME computational complexity was achieved with this method. Bit rate increases remained under 0.3 %. No results using BD measures were made available.

3.2.2.2 Prediction Unit Structure Determination

Some methods to decide or early terminate the decision of the best PU structure have also been proposed in the last few years for HEVC. The next paragraphs summarise the most relevant works in this category up to date.

In [51], the authors propose optimised schemes that conditionally evaluate certain sets of modes according to intermediate encoding decisions. The method maintains square, intra and *SKIP* PU splitting modes unchanged and proposes a conditional evaluation for *Symmetric Motion Partition* (SMP) and *Asymmetric Motion Partition* (AMP) modes, which allegedly correspond to 60 % of the computational complexity of the HM encoder (*Main* profile). Overall, complexity reductions of 51 % were achieved with the method, at the cost of a BD-rate increase of 1.3 %.

In [52], a heuristic method aims at merging $N\times N$ PUs to form larger ones instead of performing ME for every possible PU partition. The method is applied to decide all PU sizes larger than $N\times N$. When certain conditions are met, $N\times N$ partitions are merged into $2N\times N$, $N\times 2N$ or $2N\times 2N$ partitions without performing the ME operations for each one of them. Initially, ME is performed only for the four $N\times N$ partitions in a CU. If the MVs for the four partitions are identical, they are merged into one single $2N\times 2N$ PU. If the MVs are distinct, the rectangular shapes are tested in a similar manner. Experimental results point out an average complexity reduction of 34 % with an average bit rate increase of 1.37 % and a PSNR drop of 0.08 dB. No results using BD measures were made available.

In [53], a two-stage PU size decision algorithm is proposed to speed up the intra-coding process in HEVC. In the first stage, before intra-frame prediction starts, texture complexity of CTUs and its subblocks are measured in order to filter out unnecessary PU modes. The threshold for filtering PU sizes is calculated dynamically according to the content of the video sequence and to predefined coding parameters. The frame texture complexity is calculated by downsampling each 64×64 CTU to a 16×16 block and then computing its variance. In the second stage, which takes place during the intra-frame prediction, the PU sizes of neighbouring 32×32 blocks are analysed in order to skip small PU sizes. The average computational complexity reduction for the intra-coding process achieved in [53] varies from 28.8 to 44.9 %, depending on the video resolution. Average bit rate increases and PSNR decreases stayed under 0.47 % and 0.02 dB, respectively. The authors do not present results using BD measures.

An early *SKIP* mode detection scheme is proposed in [54] for complexity reduction. According to the authors, *SKIP* mode is chosen in about 83 % of CUs and detecting its occurrence would allow ignoring all the remaining modes in the RDO process. The proposed method pre-detects *SKIP* mode using the DMV and CBF information of inter $2N\times 2N$ mode. The encoder first searches the best inter $2N\times 2N$ mode (i.e. chooses between competition mode and merging mode), and after selecting the one with the minimum R-D cost, it checks the DMV and the CBF of it. If they are both equal to zero, then the best mode is determined as the *SKIP* mode and the remaining PU modes are not tested. The method reduced computational complexity by about 35 % with a BD-rate increase of 0.5 %.

Finally, [55] proposes a method in which small intra PUs are combined into larger ones, depending on the image characteristics (like texture variance), thus skipping the evaluation of certain modes. An online QP-based adaptive threshold generation is used to decide whether the smaller neighbouring PUs are to be combined in order to form a larger PU. A complexity reduction of 43.7 % was achieved with the method at the cost of an average BD-rate increase of 1.26 %.

3.2.2.3 Residual Quadtree Structure Determination

There are also some works which try to decrease the computational complexity demanded in the process of deciding the best RQT structure. However, as the computational complexity reductions achieved when constraining RQTs is very limited, as Chap. 4 will show, not many works exploit simplifications of the RQT determination procedure.

In [56] and [57], the computational complexity required to decide the RQT structure is reduced by early terminating the recursive TU splitting based on the number of nonzero transformed coefficients. The method in [56] proposes the concept of quasi-zero-blocks (QZB), which are defined by criteria based on the values of two quantities: the sum of the absolute values of coefficients and the number of nonzero coefficients. According to the authors, subtle differences between nonzero blocks and blocks with very small coefficients cannot be perceived by human eyes, so that QZBs can be used in an early termination scheme to stop splitting of TUs. An encoding time reduction of 22.8 % was achieved with the method, at the cost of a PSNR decrease of 0.04 dB. No results using BD measures were reported.

Similarly, in [57] the authors claim that the sizes chosen for the TUs and the number of nonzero coefficients are strongly correlated. Accordingly, the number of nonzero coefficients is used in this method as a threshold to stop further R-D cost evaluation of the RQT. A computational complexity decrease of 61 % in the TU processing was achieved in this method with small losses of compression efficiency. No results were reported regarding the effect on overall encoding complexity.

3.2.2.4 Other Solutions

Other solutions that reduce computational complexity by combining two or more approaches or by using a strategy different from constraining the decision of frame partitioning structures have also been proposed.

Works [58–60] optimise both the coding tree and PU structure decision processes. In [58], a complexity reduction scheme based on a method proposed in this book (presented later on, in Chap. 5) introduces a set of conditions that rely on spatial and temporal correlation to terminate early the coding tree splitting process. Simultaneously, the PU modes tested for a determined CU are chosen according to the size of neighbouring CUs. The computational complexity reduction achieved in [58] is 43 % and the BD-rate increase is around 2.2 %. The authors in [59] determine the CU depth range and the PU modes tested according to image characteristics and intermediate encoding results, such as motion homogeneity, R-D costs of neighbouring CUs and the PU mode chosen in the upper coding tree depth. Complexity was decreased by 42 % at the cost of a BD-rate increase of 1.49 %.

In [60], information from intermediate encoding steps is used to avoid exhaustive RDO searches over all possible coding tree depths and PU modes. Feature extraction to assist fast decisions is performed, and the CU size decision is performed based on a Bayesian decision rule to avoid RDO search on all possible CU sizes and PU splitting modes. Predicting the splitting of a CU is formulated as a two-class classification problem, and the features used for the decision include information such as variance of prediction error, SATD between original and predicted pixels, MV magnitude, R-D costs and others. An average complexity reduction of 41.4 % was achieved with this method, in comparison to the original HM encoder. Average BD-rate increase is around 1.88 %.

In [61], edge information from the current PU is used to reduce the number of candidate intra-frame prediction modes tested. Based on the luminance samples of each 4×4 PU, five edge strengths are calculated according to five linear filters, and the orientation of the dominant edge (i.e. the one with the largest strength) is used to choose one among five predefined sets of modes to be tested in the intra-frame prediction. In PUs larger than 4×4, the edges are still calculated for sets of 4×4 samples, and the edge with the largest number of occurrences among the sets is defined as the dominant edge of the PU. The method resulted in a reduction of 32.08 % in the intra-frame prediction complexity and an average BD-rate increase of 1.3 %.

Table 3.2 summarises the computational resource-saving strategies for HEVC surveyed in this section. Categories "CU size/depth", "PU mode" and "TU size/depth" correspond to low-complexity methods for deciding the best CU size or coding tree structure, the best PU splitting mode and the best TU size or RQT structure, respectively. Category "Other" corresponds to solutions that include more than one strategy to reduce the encoding computational complexity or that cannot be classified into any of the previous categories. The computational complexity reduction achieved with each approach is shown in the rightmost column of Table 3.2. The results are computed in comparison to the HM software in terms of encoding time reduction. Exceptions are shown as footnotes of Table 3.2.

As mentioned before in regard to Table 3.1, it is also important to notice that most of the approaches presented in Table 3.2 cannot be directly compared with one another due to different encoding conditions. However, the table provides useful information for identifying the best computational resource-saving strategies for HEVC published so far. By comparing the results of Tables 3.1 and 3.2, it is possible to conclude that the research focusing on approaches for HEVC still have

Table 3.2 Computational complexity reduction strategies for HEVC

Category	Approach	Reference	Complexity reduction (%)
CU size/depth	Fast splitting for intra-coding	[43]	50
	Top skip+early termination	[44]	40
	Early termination	[45]	48
	Early termination	[46]	38
	Motion divergence	[47]	43
	Temporal correlation in neighbouring frames	[48]	20–33
	Spatial/temporal correlation	[49]	25
	Skip rarely used tree depths	[50]	45[a]
PU mode	Conditional evaluation	[51]	51
	Merge smaller PUs into larger PUs	[52]	34
	Filter out unnecessary intra PU sizes	[53]	28.8–44.9[b]
	Early *SKIP* mode detection	[54]	35
	Small intra PUs combined into larger PUs	[55]	44
TU size/depth	QZB-based early termination	[56]	23
	Nonzero coefficient-based early termination	[57]	61[c]
Other	Spatial/temporal correlation	[58]	43
	Image characteristics and intermediate results	[59]	42
	Feature extraction	[60]	41.4
	Edge information for intra mode decision	[61]	32.08[b]

[a]Time reduction for the ME module only
[b]Time reduction for the intra-frame MD only
[c]Time reduction for the RQT processing only

room for improvement to identify whether and how complexity reduction levels similar to those of the best performing methods developed for H.264/AVC can be achieved. This is the main motivation for the research work presented later in Chap. 6.

3.3 Computational Complexity Scaling

To efficiently manage computational complexity, it is not enough to implement low-complexity methods. Since these methods usually incur in compression efficiency losses, they must be wisely and gradually employed in order to reduce computational complexity up to a certain desired target, avoiding unnecessary losses in terms of R-D efficiency. This is usually achieved by designing scaling systems in which computational complexity can be adaptively adjusted according to specific conditions, such as the device's battery status, time limitations imposed by the transmission environment and user preferences. Therefore, such systems generally have multiple operation modes, which can be dynamically chosen by sensing environment changes or even by a user's preference.

In the last years, dynamic scaling of computational complexity in video encoding has been a very active research field. The ideal solution for a complexity management system would be to extend the original RDO problem to a third dimension such as rate-distortion-complexity optimisation (RDCO) or power-rate-distortion optimisation (PRDO). This solution would find the best encoder configuration leading to the optimal visual quality under predefined rate and computational complexity constraints. However, joint rate-distortion-complexity (R-D-C) analysis is an extremely complex task. Indeed, the lack of analytic models that relate rate, distortion and complexity prevents analytic solutions and the empirical solutions are not practical, as they require trying a huge number of possible combinations of modes and encoding parameters. As exhaustive search is usually infeasible in complexity-constrained environments, several heuristic solutions have been proposed to dynamically adjust the computational complexity of video encoding in order to manage the use of available computational resources.

3.3.1 Computational Complexity Scaling for H.264/AVC

Due to its high computational complexity, the ME process is used by many methods to scale the encoding computational complexity of H.264/AVC and other previous standards. A complexity scaling scheme for the MPEG-4 encoder was presented in [62] based on adjusting ME parameters and using the search options of FS, the TSS and the *Spiral Search* algorithms. The remaining parameters are the SA size, the SAD threshold for early termination of the *Spiral Search* algorithm, the use of pixel subsampling and the number of bits used to represent a pixel.

A classification-based method is proposed in [63] for the ME process. MBs are classified into different categories according to their importance in the frame, and a complexity-controlled ME scheme applies different operations to each MB according to its class. Initially, a total computation budget for the video sequence is divided into computation allowances for each frame. Then, the frame budget is divided into three independent sub-budgets, which are assigned to each class of MBs. When performing ME, each frame is classified into one of the three classes, and a computation budget is allocated to each MB according to its class. Finally, according to the MB computation budget, the encoder uses more or less computation to estimate the motion of that MB.

Three other ME adjusting parameters are proposed in [6]: partial cost evaluation for fractional motion estimation (FRME), block size adjustment for FRME and search range adjustment for integer motion estimation (IME). By combining these parameters, 12 configurations presenting different trade-off between compression efficiency and computational complexity were defined and used. Based on the complexity measure from the previous frames, the complexity scaling algorithm allocates a certain budget for each group of MBs in an image and then for each MB within the group.

A two-stage complexity scaling strategy is proposed in [8] based on adjusting ME. In the first stage, an encoding time scaling algorithm is applied. It consists of encoding the whole frame only using the inter 16×16 mode and then, based on encoding time information for this mode, estimating the total encoding time, the target encoding time and the parameters to be used in the second stage for complexity scaling. In the second stage, the number of *SKIP* MBs in the frame is used to adjust the encoder computational complexity to a determined target complexity level.

As in other methods aiming to reduce the computational complexity, many complexity scaling methods are based on the adjustment of both MD and ME operations. In [64] a complexity scaling scheme is proposed based on adjusting parameters that affect the aggressiveness of an early stop criterion for ME, the number of prediction modes tested in the MD and the accuracy of ME steps for inter modes. Before deciding the number of modes to test, they are ordered based on the statistical frequency of the optimal modes for a given type of video, so that the first modes tested are those which most probably yield the best R-D performance. The computational complexity is scaled by adjusting one single parameter, which is mapped to the algorithmic parameters based on a rule tuned by an offline training process that uses several typical video sequences.

A similar approach based on sorting modes was also used in [65] to scale the MD computational complexity. A ranking with the most popular ones, including intra and inter modes, compose a subset which is tested in the RDO process, while the remaining modes are suppressed from the tests. Initially, a few MBs are randomly selected for the frequency distribution analysis. Each mode is then associated with a frequency of occurrence and a computational complexity. Then, based on the target complexity to encode a complete image, the dominant mode set is chosen and used to encode the next frame.

In [66], the computational complexity scaling is divided into two problems: how to allocate the available computational resources to different frames and encoding modules and how to optimally use the allocated resources by adjusting the encoding parameters. To solve the first problem, a computation allocation model is proposed to distribute the available resources among the video frames. The second problem is solved by using a complexity-adjustable ME and a complexity-adjustable MD. The ME complexity is adjusted by allowing more or less operations (such as FRME, searching point refinement, etc.) to be executed, while the MD complexity is adjusted by allowing more or less modes to be tested in the RDO process. The list of tested modes is sorted according to their occurrence frequency in the spatial and temporal neighbouring MBs.

The MD complexity was scaled in [9, 67] by an adaptive method at the MB level. The computational cost of ME in 16×16 blocks is taken as the basic unit, and the computational costs for the remaining modes were obtained through empirical simulations and represented as weighting factors of the basic unit. A target complexity is calculated based on the total computational complexity estimated for a frame, which depends on the number of MBs in the frame and on the computational weighting factors previously defined. The computational budget allocated for each MB is calculated based on the target complexity, on the complexity consumed by the previously encoded MBs in the frame and on the number of MBs already coded. The computational budget for a determined MB is then used to adaptively choose which modes are to be tested and which are not.

A MD early termination is used in [68] to decrease the encoding process computational complexity. By calculating the difference between the cost of encoding a MB as *SKIP* and an estimated coding cost, the encoder is able to stop the MD evaluation process just after encoding the MB as *SKIP*. A threshold calculated using conditional probability estimates of skipped and not skipped MBs is used in the early termination decision. The authors also propose a complexity scaling method which aims at maintaining a target level of complexity through a feedback algorithm that updates probability models to reduce R-D performance losses.

Other parameters which scale the complexity of other operations besides ME and MD are explored in some works. In [69], an empirical study on the controllability of parameters for complexity scaling on video encoding cloud services is presented. The work presents experimental results in terms of encoding time, bit rate and objective quality when varying the number of B frames, the level of refinement for subpixel ME and the operations performed in quantisation.

In [70], the number of reference frames, the method used for subpixel ME, the partition sizes allowed for intra-frame and inter-frame prediction and the quantisation approach are used as the parameters to adjust computational complexity. Considering all possible combinations among these four parameters would result in a total of 3360 possible parameter settings, a number so large that makes searching the best configuration infeasible in real-time applications. Two fast algorithms are proposed for finding the parameter settings which leads to high distortion-complexity performance. The algorithms are based on the *Generalised Breiman, Friedman, Olshen and Stone* (GBFOS) Algorithm [71] and use training sequences to find the best parameter settings.

The number of motion search positions and the frame rate were the two parameters used in the method proposed in [72]. By using the *adaptive critic design* technique [73], a class of approximate dynamic programming methods, an online complexity scaling scheme was developed based on neural networks.

A two-level method is proposed in [74]. In the frame-level algorithm, the encoding process is not changed in order to maintain acceptable image quality, but frames are dropped when necessary to decrease the amount of computations. In the per-frame algorithm, computational complexity is scaled for each frame in order to achieve the target encoding time. The frame-level algorithm calculates a target encoding time for each frame in the video sequence based on the total delay experienced by the frame in the input buffer. The target time is then used by the per-frame algorithm, which adjusts computational complexity in a frame by increasing or decreasing the number of MBs encoded as *SKIP*, similarly to [68].

A dynamic framework which consists of a set of optimised core components is proposed in [75]. The ME, DCT, quantisation and MD processes can be configured to achieve a desired computation-performance trade-off in the encoder. These modules can all be assembled to form an H.264/AVC encoder with various degrees of computational complexity, which is able to adapt itself according to the available computational resources. Eleven parameters are used to adjust the computational effort of different encoder modules, such as the number of ME search points and whether or not DCT is applied to the residual MB. As determining the best combination of parameters for each video through exhaustive optimisation is quite computationally demanding, a simpler, suboptimal greedy optimisation method is used.

RDCO and other similar approaches have also been proposed in several works, which treat the problem as an extension of RDO. In [7] and [76], a power-rate-distortion (PRD) analysis framework was developed in order to build a parametric video encoding architecture which scales the computational complexity of its modules by varying encoding parameters. Based on the R-D behaviour of these parameters and on their associated computational complexity, a PRD model was built and used to determine the best configuration of parameters according to the available power supply level of the device in which the encoding is done and on the target bit rate. The same authors propose in [77] an operational approach for offline PRD analysis and modelling based on a wide set of training data. Based on the models developed, a control database for online resource allocation and energy minimisation is proposed.

Game theoretical analysis is used in [78] to model the power consumption in video encoders. The encoder is divided into modules, which are treated as players competing for the use of a computational resource on a limited budget aiming at maximising efficiency. In [79], the complexity dimension was added to the RDO strategy. For each particular encoder set-up, the total bit rate (R), PSNR (D) and ratio between the time spent to encode a training sequence and the time spent by the full-featured encoder (C) are calculated. The RDC points are then projected into a 2D set of points (lying in the D-C plane, for a given constant bit rate), and a lookup table is built from the points in the convex hull of the set in order to provide optimal starting RDC points. The trellis quantisation, the level of refinement in ME and the number of partitioning modes tested are the adjusted parameters.

Table 3.3 summarises the strategies for computational resource management proposed for use in H.264/AVC. Categories "ME" and "MD" correspond, respectively, to low-complexity ME and MD methods. Category "FR" corresponds to methods which change the output frame rate (i.e. discarding frames) in order to scale the encoding computational complexity. Categories "RDCO" and "Combined" correspond to rate-distortion-complexity optimisation and combined strategies. The computational complexity reductions achieved are generally computed with reference to the H.264/AVC model encoder in terms of encoding time. Exceptions are shown as footnotes of Table 3.3.

Table 3.3 Computational complexity scaling strategies for H.264/AVC

Category	Approach	Reference	Complexity reduction
ME	Adjusting ME parameters and algorithms	[62]	Up to 40 %[a]
	MB importance classification	[63]	Up to 60 %
	Adjusting ME parameters	[6]	–
	Encoding time estimation	[8]	Up to 70 %
MD	Mode ranking	[65]	Up to 70 %
	SKIP early termination	[68]	Up to 50 %
ME, MD	Adjusting ME parameters and modes	[64]	Up to 80 %
	Frame-level resource allocation	[66]	Up to 95 %[b]
	MB-level resource allocation	[9, 67]	Up to 91.2 %
ME, FR	ME and frame rate (FR) adjustment	[72]	Up to 90 %
MD, FR	Frame-level and MB-level resource allocation	[74]	Up to 50 %
RDCO	Power-rate-distortion model	[76]	Up to 78.6 %[a]
	Game theoretical analysis	[78]	Up to 95 %[a]
	ME, MD and quantisation adjustment	[79]	Up to 85 %
Combined	ME, MD, quantisation adjustment	[69]	–
	ME, MD and quantisation adjustment	[70]	Up to 94.1 %
	ME, MD, transform and quantisation adjustment	[75]	Up to 50 %

[a]Power consumption reduction
[b]Uses a computational complexity measure based on the SAD computation cost

Although most of the approaches presented in Table 3.3 cannot be directly compared with one another due to different encoding conditions, the results can still serve as guidelines for identifying the best computational resource management strategies.

3.3.2 *Computational Complexity Scaling in HEVC*

Even though several solutions for complexity scaling in H.264/AVC have been proposed, not so many strategies have been published for HEVC so far. This is the main focus of Chap. 5 of this book, which presents a set of computational complexity scaling strategies for HEVC.

The authors in [80] present a complexity scaling scheme based on a contribution detailed later in Chap. 5. Their approach allows defining maximum coding tree depth values independently for each encoded frame, so that the computational complexity can be adjusted according to a given complexity budget. The initial frames of a video are normally encoded and then a game-theoretic approach is used to choose the depth values for the N next frames to be encoded. The maximum complexity reduction achieved by the method is around 40 % with an average PSNR loss of 0.027 dB in such case. No results were made available using BD measures.

Table 3.4 Computational complexity scaling strategies for HEVC

Category	Approach	Reference	Complexity reduction
CU size/depth	Game-theoretic R-D-C optimisation	[80]	Up to 40 %
PU mode	Mode mapping-based mode selection	[81]	Up to 50 %

In [81], a complexity scaling scheme allows adjusting the number of evaluated PU splitting modes for inter-predicted CUs according to a target computational complexity. Through investigations on the MVs correlations, the authors propose a mode mapping method for PU splitting mode selection. Linear programming strategies are used to allocate the computational complexity and adjust the number of candidate modes. The maximum complexity reduction achieved by the method is around 50 % with an average BD-rate increase of 5.9 % in such case.

Table 3.4 summarises the two computational complexity scaling strategies for HEVC. The maximum computational complexity reduction achieved with each approach is shown in the rightmost column of the table. Results are computed in comparison to the HM software in terms of encoding time reduction. By comparing Tables 3.3 and 3.4, it is possible to conclude that the research focusing on approaches for complexity scaling of HEVC is still in its first steps. While several works have been proposed for H.264/AVC, only a couple of strategies are available for HEVC so far. This is the main motivation for the research work presented later in Chap. 5.

3.4 Challenges and Conclusions

As explained in the previous sections, research on computational complexity management for video encoding through reduction and scaling of algorithm complexity made significant advances in recent years. However, the fast evolution of electronic devices in terms of computational power and display technologies and the development of new and more complex video coding standards, like HEVC, introduce new important challenges to be solved.

The challenges come from different sides. The increasing screen resolution of current multimedia-capable electronic devices and cameras allows higher-resolution video sequences to be played, recorded and transmitted. Video content with higher resolutions require greater computational efforts to be processed and transmitted, increasing the power consumption in such devices. At the same time, such high-resolution screens require more energy to work, limiting even more the energy available for computational operations. In order to reduce the number of bits and decrease the energy spent on transmission, more efficient compression methods must be used, which in turn increases even more the computational complexity and power consumption. It is easy to perceive that there is no way to avoid or ignore the computational complexity increase incurred by the latest technology advances.

In addition, new video communication paradigms have arisen recently in the form of wireless visual sensor networks (VSN), inter-vehicle communication networks with video transmission support and other ad hoc networks which permit hop-to-hop video transmission or video diffusion through peer-to-peer (P2P) overlays. Since in many cases the nodes of these heterogeneous networks are mobile devices with limited energy and computation resources, the support of video encoding requires the use of carefully crafted domain-specific computational resource management techniques to ensure longer autonomies without significant encoded video quality degradations.

As discussed in the previous sections, many parameters of current generation encoders can be varied to reduce and scale the use of computation and energy resources in multimedia systems. This large number of parameters makes the analysis of the encoder's R-D-C efficiency a very complex task but also allows finding better ways of reducing and scaling computational complexity. It is worth noting, however, that the HEVC reference software includes less encoding parameters than H.264/AVC, which on one side simplifies the analysis of the encoder R-D-C surface but on the other side reduces the number of encoding possibilities to be considered when developing a complexity scaling system. This is a challenge to be overcome by identifying in the HEVC standard which tasks should be parameterised in order to allow complexity scaling at a finer grain. Also, as different video sequences with different contents have different R-D-C characteristics, developing efficient and accurate content-aware models is also another important open issue to be solved.

This chapter has presented a study on the main works published so far, focusing on computational complexity modelling, reduction and scaling for H.264/AVC and HEVC. Most of the presented approaches focus on the computational complexity demanded by the ME and the MD processes of H.264/AVC, but some works aiming at complexity reduction have already been exploited for HEVC. However, such methods for HEVC are still rare and do not achieve complexity scalability levels as large as the methods designed for H.264/AVC.

When comparing the tables presented in previous sections, it becomes clear that, due to the intrinsic lower computational complexity of H.264/AVC in comparison to HEVC and due to the fact that it has been available for a longer period for researchers, engineers and system designers, the approaches focusing on the former encoder achieve much higher computational complexity reduction levels than those focusing on the latter. This is the main motivation for the research work presented in the next chapters of this book.

References

1. M. Mattavelli, S. Brunetton, Implementing real-time video decoding on multimedia processors by complexity prediction techniques. IEEE Trans. Consum. Electron. **44**, 760–767 (1998)
2. J. Valentim, P. Nunes, F. Pereira, Evaluating MPEG-4 video decoding complexity for an alternative video complexity verifier model. IEEE Trans. Circ. Syst. Video Technol. **12**, 1034–1044 (2002)
3. M. van der Schaar, Y. Andreopoulos, Rate-distortion-complexity modeling for network and receiver aware adaptation. IEEE Trans. Multimed. **7**, 471–479 (2005)
4. Y. Wang, C. Shih-Fu, Complexity adaptive H.264 encoding for light weight streams, in *2006 IEEE International Conference on Acoustics, Speech and Signal Processing* (2006), pp. II–II.

5. L. Szu-Wei, C.C.J. Kuo, Motion compensation complexity model for decoder-friendly H.264 system design, in *2007 IEEE Workshop on Multimedia Signal Processing* (2007), pp. 119–122.
6. C.E. Rhee, J.S. Jung, H.J. Lee, A real-time H.264/AVC encoder with complexity-aware time allocation. IEEE Trans. Circ. Syst. Video Technol. **20**, 1848–1862 (2010)
7. H. Zhihai, L. Yongfang, C. Lulin, I. Ahmad, W. Dapeng, Power-rate-distortion analysis for wireless video communication under energy constraints. IEEE Trans. Circ. Syst. Video Technol. **15**, 645–658 (2005)
8. W. Kim, J. You, J. Jeong, Complexity control strategy for real-time H.264/AVC encoder. IEEE Trans. Consum. Electron. **56**, 1137–1143 (2010)
9. L. Xiang, M. Wien, J.R. Ohm, Rate-complexity-distortion optimization for hybrid video coding. IEEE Trans. Circ. Syst. Video Technol. **21**, 957–970 (2011)
10. M. Zhan, H. Hao, W. Yao, On complexity modeling of H.264/AVC video decoding and its application for energy efficient decoding. IEEE Trans. Multimed. **13**, 1240–1255 (2011)
11. R. Garcia, D. Ruiz Coll, H. Kalva, G. Fernandez-Escribano, HEVC decision optimization for low bandwidth in video conferencing applications in mobile environments, in *IEEE International Conference on Multimedia and Expo (ICME 2013)*, San Jose, USA (2013).
12. A. Yong-Jo, H. Tae-Jin, S. Dong-Gyu, H. Woo-Jin, Complexity model based load-balancing algorithm for parallel tools of HEVC, in *Visual Communications and Image Processing (VCIP), 2013* (2013), pp. 1–5.
13. T. Koga, K. Iinuma, A. Hirano, Y. Iijima, T. Ishiguro, Motion compensated interframe coding for video conferencing, in *National Telecommunications Conference*, New Orleans (1981), pp. G5.3.1–G5.3.5.
14. L. Lurng-Kuo, E. Feig, A block-based gradient descent search algorithm for block motion estimation in video coding. IEEE Trans. Circ. Syst. Video Technol. **6**, 419–422 (1996)
15. L. Renxiang, Z. Bing, M.L. Liou, A new three-step search algorithm for block motion estimation. IEEE Trans. Circ. Syst. Video Technol. **4**, 438–442 (1994)
16. T. Jo Yew, R. Surendra, M. Ranganath, A.A. Kassim, A novel unrestricted center-biased diamond search algorithm for block motion estimation. IEEE Trans. Circ. Syst. Video Technol. **8**, 369–377 (1998)
17. Z. Shan, M. Kai-Kuang, A new diamond search algorithm for fast block-matching motion estimation. IEEE Trans. Image Process. **9**, 287–290 (2000)
18. Z. Ce, L. Xiao, L.P. Chau, Hexagon-based search pattern for fast block motion estimation. IEEE Trans. Circ. Syst. Video Technol. **12**, 349–355 (2002)
19. I.E. Richardson, *Video Codec Design: Developing Image and Video Compression Systems* (Wiley, New York, 2002)
20. B. Xuan-Quang, T. Yap-Peng, Adaptive dual-cross search algorithm for block-matching motion estimation. IEEE Trans. Consum. Electron. **50**, 766–775 (2004)
21. Y.-L.S. Lin, C.-Y. Kao, H.-C. Kuo, J.-W. Chen, *VLSI Design for Video Coding: H.264/AVC Encoding from Standard Specification to Chip* (Springer Publishing Company, Incorporated, New York, 2010)
22. S. Eckart, C.E. Fogg, ISO-IEC MPEG-2 software video codec, in *SPIE Conference Visual Communications and Image Processing* (1995), pp. 100–109.
23. C. Chok-Kwan, P. Lai-Man, Normalized partial distortion search algorithm for block motion estimation. IEEE Trans. Circ. Syst. Video Technol. **10**, 417–422 (2000)
24. K. Lengwehasarit, A. Ortega, Probabilistic partial-distance fast matching algorithms for motion estimation. IEEE Trans. Circ. Syst. Video Technol. **11**, 139–152 (2001)
25. W. Li, E. Salari, Successive elimination algorithm for motion estimation. IEEE Trans. Image Process. **4**, 105–107 (1995)
26. M. Moecke, R. Seara, Sorting rates in video encoding process for complexity reduction. IEEE Trans. Circ. Syst. Video Technol. **20**, 88–101 (2010)
27. S. Saponara, M. Casula, F. Rovati, D. Alfonso, L. Fanucci, Dynamic control of motion estimation search parameters for low complex H.264 video coding. IEEE Trans. Consum. Electron. **52**, 232–239 (2006)

28. M. Bystrom, I. Richardson, Y. Zhao, Efficient mode selection for H.264 complexity reduction in a Bayesian framework. Signal Process. Image Commun. **23**, 71–86 (2008)
29. L.-J. Pan, H. Yo-Sung, A fast mode decision algorithm for H.264/AVC intra prediction, in *2007 IEEE Workshop on Signal Processing Systems* (2007), pp. 704–709.
30. C. Chun-Hao, C. Jia-Wei, C. Hsiu-Cheng, Y. Yao-Chang, W. Jinn-Shyan, G. Jiun-In, A quality scalable H.264/AVC baseline intra encoder for high definition video applications, in *2007 IEEE Workshop on Signal Processing Systems* (2007), pp. 521–526.
31. L. De-Wei, K. Chun-Wei, C. Chao-Chung, L. Yu-Kun, C. Tian-Sheuan, A 61MHz 72K Gates 1280x720 30FPS H.264 intra encoder, in *2007 IEEE International Conference on Acoustics, Speech and Signal Processing* (2007), pp. II-801–II-804.
32. T. An-Chao, W. Jhing-Fa, Y. Jar-Ferr, L. Wei-Guang, Effective Subblock-based and pixel-based fast direction detections for H.264 Intra prediction. IEEE Trans. Circ. Syst. Video Technol. **18**, 975–982 (2008)
33. L. Yu-Ming, S. Yu-Ting, L. Yinyi, SATD-based intra mode decision for H.264/AVC video coding. IEEE Trans. Circ. Syst. Video Technol. **20**, 463–469 (2010)
34. K. Hyungjoon, Y. Altunhasak, Low-complexity macroblock mode selection for H.264-AVC encoders, in *2004 International Conference on Image Processing*, vol. 2. (2004), pp. 765–768.
35. F. Wang, Y. Fan, Y. Lan, W. Liu, Fast intra mode decision algorithm in H.264/AVC using characteristics of transformed coefficients, in *2008 International Conference on Visual Information Engineering* (2008), pp. 245–249
36. L. Yu-Ming, W. Jyun-De, L. Yinyi, An improved SATD-based intra mode decision algorithm for H.264/AVC, in *2009 IEEE International Conference on Acoustics, Speech and Signal Processing* (2009), pp. 1029–1032.
37. G. Correa, C. Diniz, S. Bampi, D. Palomino, R. Porto, L. Agostini, Homogeneity and distortion-based intra mode decision architecture for H.264/AVC, in *2010 IEEE International Conference on Electronics, Circuits, and Systems* (2010), pp. 591–594.
38. G. Correa, D. Palomino, C. Diniz, L. Agostini, S. Bampi, SHBS: a heuristic for fast inter mode decision of H.264/AVC standard targeting VLSI design, in *2011 IEEE International Conference on Multimedia and Expo* (2011), pp. 1–4.
39. G. Correa, D. Palomino, C. Diniz, S. Bampi, L. Agostini, Low-complexity hierarchical mode decision algorithms targeting VLSI architecture design for the H.264/AVC video encoder. VLSI Des. **2012**, 20 (2012)
40. L. Jeyun, J. Byeungwoo, Fast mode decision for H.264, in *2004 IEEE International Conference on Multimedia and Expo*, vol. 2 (2004), pp. 1131–1134.
41. D. Wu, F. Pan, K.P. Lim, S. Wu, Z.G. Li, X. Lin, S. Rahardja, C.C. Ko, Fast intermode decision in H.264/AVC video coding. IEEE Trans. Circ. Syst. Video Technol. **15**, 953–958 (2005)
42. S. Liquan, L. Zhi, Z. Zhaoyang, S. Xuli, Fast inter mode decision using spatial property of motion field. IEEE Trans. Multimed. **10**, 1208–1214 (2008)
43. C. Seunghyun, K. Munchurl, Fast CU splitting and pruning for suboptimal CU partitioning in HEVC intra coding. IEEE Trans. Circ. Syst. Video Technol. **23**, 1555–1564 (2013)
44. M.B. Cassa, M. Naccari, F. Pereira, Fast rate distortion optimization for the emerging HEVC standard, in *2012 Picture Coding Symposium* (2012), pp. 493–496.
45. L. Jong-Hyeok, P. Chan-Seob, K. Byung-Gyu, Fast coding algorithm based on adaptive coding depth range selection for HEVC, in *2012 IEEE International Conference on Consumer Electronics - Berlin (ICCE-Berlin)* (2012), pp. 31–33.
46. K. Goswami, B.-G. Kim, D.-S. Jun, S.-H. Jung, J.S. Choi, Early coding unit (CU) splitting termination algorithm for high efficiency video coding (HEVC). ETRI J. **36**(3), 407–417 (2014)
47. X. Jian, L. Hongliang, W. Qingbo, M. Fanman, A fast HEVC inter CU selection method based on pyramid motion divergence. IEEE Trans. Multimed. **16**, 559–564 (2014)
48. S.-C. Tai, C.-Y. Chang, B.-J. Chen, J.-F. Hu, Speeding up the decisions of quad-tree structures and coding modes for HEVC coding units, in *Advances in intelligent systems and applications - Volume 2*, ed. by J.-S. Pan, C.-N. Yang, C.-C. Lin, vol. 21 (Springer, Berlin, 2013), pp. 393–401
49. Y. Zhang, H. Wang, Z. Li, Fast coding unit depth decision algorithm for interframe coding in HEVC, in *2013 Data Compression Conference*, Snowbird, Utah (2013), pp. 53–62.

50. L. Jie, S. Lei, T. Ikenaga, S. Sakaida, Content based hierarchical fast coding unit decision algorithm for HEVC, in *2011 International Conference on Multimedia and Signal Processing* (2011), pp. 56–59.
51. J. Vanne, M. Viitanen, T. Hamalainen, Efficient mode decision schemes for HEVC inter prediction. IEEE Trans. Circ. Syst. Video Technol. **24**, 1579–1593 (2014)
52. F. Sampaio, S. Bampi, M. Grellert, L. Agostini, J. Mattos, Motion vectors merging: low complexity prediction unit decision heuristic for the inter-prediction of HEVC encoders, in *2012 IEEE International Conference on Multimedia and Expo* (2012), pp. 657–662.
53. T. Guifen, S. Goto, Content adaptive prediction unit size decision algorithm for HEVC intra coding, in *2012 Picture Coding Symposium* (2012), pp. 405–408.
54. K. Jaehwan, Y. Jungyoup, W. Kwanghyun, J. Byeungwoo, Early determination of mode decision for HEVC, in *2012 Picture Coding Symposium* (2012), pp. 449–452.
55. M.U.K. Khan, M. Shafique, J. Henkel, An adaptive complexity reduction scheme with fast prediction unit decision for HEVC intra encoding, in *IEEE International Conference on Image Processing (ICIP)* (2013).
56. Y. Shi, Z. Gao, X. Zhang, Early TU split termination in HEVC based on Quasi-Zero-Block, in *3rd International Conference on Electric and Electronics* (2013).
57. C. Kiho, E.S. Jang, Early TU decision method for fast video encoding in high efficiency video coding. Electron. Lett. **48**, 689–691 (2012)
58. H. Wei-Jhe, H. Hsueh-Ming, Fast coding unit decision algorithm for HEVC, in *Signal and Information Processing Association Annual Summit and Conference (APSIPA), 2013 Asia-Pacific* (2013), pp. 1–5.
59. S. Liquan, L. Zhi, Z. Xinpeng, Z. Wenqiang, Z. Zhaoyang, An effective CU size decision method for HEVC encoders. IEEE Trans. Multimed. **15**, 465–470 (2013)
60. S. Xiaolin, Y. Lu, C. Jie, Fast coding unit size selection for HEVC based on Bayesian decision rule, in *2012 Picture Coding Symposium* (2012), pp. 453–456.
61. T.L. da Silva, L.V. Agostini, L.A. da Silva Cruz, Fast HEVC intra prediction mode decision based on EDGE direction information, in *Signal Processing Conference (EUSIPCO), 2012 Proceedings of the 20th European* (2012), pp. 1214–1218.
62. P. Jain, A. Laffely, W. Burleson, R. Tessier, D. Goeckel, Dynamically parameterized algorithms and architectures to exploit signal variations. J VLSI Sig. Process. **36**, 27–40 (2004)
63. L. Weiyao, K. Panusopone, D.M. Baylon, S. Ming-Ting, A computation control motion estimation method for complexity-scalable video coding. IEEE Trans. Circ. Syst. Video Technol. **20**, 1533–1543 (2010)
64. E. Akyol, D. Mukherjee, L. Yuxin, Complexity control for real-time video coding, in *2007 IEEE International Conference on Image Processing* (2007), pp. I-77–I-80.
65. T. A. da Fonseca, R.L. de Queiroz, Macroblock sampling and mode ranking for complexity scalability in mobile H.264 video coding, in *2009 IEEE International Conference on Image Processing* (2009), pp. 3753–3756.
66. S. Li, L. Yan, W. Feng, L. Shipeng, G. Wen, Complexity-constrained H.264 video encoding. IEEE Trans. Circ. Syst. Video Technol. **19**, 477–490 (2009)
67. L. Xiang, M. Wien, J. R. Ohm, Medium-granularity computational complexity control for H.264/AVC, in *2010 Picture Coding Symposium* (2010), pp. 214–217.
68. C.S. Kannangara, I.E. Richardson, M. Bystrom, Z. Yafan, Complexity control of H.264/AVC based on mode-conditional cost probability distributions. IEEE Trans. Multimed. **11**, 433–442 (2009)
69. X. Li, Y. Cui, Y. Xue, Towards an automatic parameter-tuning framework for cost optimization on video encoding cloud. Int. J. Digit. Multimed. Broadcast. **2012**, 1–11 (2012).
70. R. Vanam, E. Riskin, R. Ladner, S. Hemami, Fast algorithms for designing nearly optimal lookup tables for complexity control of the H.264 encoder. SIViP **7**, 991–1003 (2013)
71. L. Breiman, *Classification and regression trees* (Chapman & Hall, New York, 1984)
72. Z. Sun, C. Xi, H. Zhihai, Adaptive critic design for energy minimization of portable video communication devices, in *2008 International Conference on Computer Communications and Networks* (2008), pp. 1–5.

73. D.V. Prokhorov, D.C. Wunsch, Adaptive critic designs. IEEE Trans. Neural Netw. **8**, 997–1007 (1997)
74. C.S. Kannangara, I.E. Richardson, A.J. Miller, Computational complexity management of a real-time H.264/AVC encoder. IEEE Trans. Circ. Syst. Video Technol. **18**, 1191–1200 (2008)
75. I.R. Ismaeil, A. Docef, F. Kossentini, R.K. Ward, A computation-distortion optimized framework for efficient DCT-based video coding. IEEE Trans. Multimed. **3**, 298–310 (2001)
76. L. Yongfang, I. Ahmad, Power and distortion optimization for pervasive video coding. IEEE Trans. Circ. Syst. Video Technol. **19**, 1436–1447 (2009)
77. H. Zhihai, C. Wenye, C. Xi, Energy minimization of portable video communication devices based on power-rate-distortion optimization. IEEE Trans. Circ. Syst. Video Technol. **18**, 596–608 (2008)
78. W. Ji, J. Liu, M. Chen, Y. Chen, Power-efficient video encoding on resource-limited systems: a game-theoretic approach. Future Gener. Comput. Syst. **28**, 427–436 (2012)
79. T.A. da Fonseca, R.L. de Queiroz, Complexity-constrained rate-distortion optimization for H.264/AVC video coding, in *2011 IEEE International Symposium on Circuits and Systems*, Rio de Janeiro, (2011).
80. A. Ukhanova, S. Milani, S. Forchhammer, Game-theoretic rate-distortion-complexity optimization for HEVC, in *2013 IEEE International Conference on Image Processing*, Melbourne, Australia, (2013), pp. 1995–1999.
81. T. Zhao, Z. Wang, S. Kwong, Flexible mode selection and complexity allocation in high efficiency video coding. IEEE J. Sel. Top. Signal. Process. **7**, 1135–1144 (2013)

Chapter 4
Performance and Computational Complexity Assessment of HEVC

This chapter presents an experimental study performed with the research goal of characterising and evaluating the behaviour and performance of the HEVC encoder. A first set of experiments was carried out in order to identify which tools most affect the HEVC encoding process in terms of encoding efficiency and computational complexity [1]. Then, a second set of experiments was defined to analyse the impact of using different frame partitioning structures in both the compression ratio and encoding computational complexity.

The results of this study provided relevant insight for developing the complexity reduction and scaling methods presented in the following chapters of this book.

4.1 Analysis of HEVC Encoding Tools

As explained in Chap. 2, the HEVC encoder includes a large number of tools, each one having a different contribution to the overall encoding efficiency and complexity. The operation of each tool and their functional modes are determined by configuration parameters that may be set to several different values during the encoding process.

This section presents an analysis on the impact in both the R-D efficiency and the encoding computational complexity when enabling and disabling each tool, as well as employing different parameter values whenever possible. By doing so, it is possible to identify which tools present the best trade-off between compression efficiency and computational complexity and thus should have enabling priority in a complexity-constrained system.

G. Corrêa et al., *Complexity-Aware High Efficiency Video Coding*,
DOI 10.1007/978-3-319-25778-5_4

4.1.1 Experimental Setup and Methodology

The methodology defined for the experimental study presented in this section comprised two main steps. Firstly, the coding tools that present stronger impact in both coding efficiency and computational complexity were identified. To that end, the individual contribution of each and every encoding tool to such performance indicators was experimentally evaluated. Secondly, those tools identified in the first step and ordered according to encoding gain per complexity increase were selected for further analysis, which consisted in evaluating the impact of these tools when enabled in a cumulative sequence (i.e. first enabling tool *A* and then tools *A* and *B* and so on).

To conduct the experiments, 12 video sequences that differ broadly from one another in terms of frame rate, bit depth, motion and texture characteristics as well as spatial resolution were used. The 12 selected sequences are the *BlowingBubbles*, *RaceHorses1*, *BasketballDrillText*, *PartyScene*, *BQMall*, *SlideShow*, *vidyo1*, *vidyo4*, *ParkScene*, *BasketballDrive*, *NebutaFestival* and *Traffic*, which are all detailed in the Appendix A of this book.

The reference software used was the HM version 7.0 (HM7) [2], which was compiled using *Microsoft Visual Studio C++ Compiler* under the *Release* compilation mode (i.e. allowing compiling optimisations). All tests were performed in a single core of a clustered computer based on *Intel® Xeon® E5520* (2.27 GHz) processors running the *Windows Server 2008 HPC* operating system. The computational complexity was measured in terms of processing time, reported by the *Intel® VTune™ Amplifier XE* software profiler [3]. The *low delay* temporal configuration was used in all the tests.

4.1.2 Identification of Relevant Parameters

As revealed by experiments carried out in the scope of this chapter, as well as in other works conducted by the JCT-VC group, the impact of different HEVC tools on encoding efficiency and computational complexity is highly variable [4]. Since testing all possible combinations of coding tools and functional modes for all video sequences would require an inordinate amount of time to the point of being unfeasible and produce a huge amount of data, a preliminary exploration was first conducted to identify the configuration parameters that would be more important to this study. These exploratory experiments were performed by varying one parameter at a time, in the multidimensional encoder configuration parameter set, such that the impact of enabling each tool could be separately analysed. Starting with a baseline encoder configuration, each tool was enabled (and then reset to the default value), one after the other, and the resulting image quality, bit rate and encoding computational complexity were recorded for comparison with the reference baseline configuration. In each case, including the baseline configuration, the encoding structure was set to support CUs of up to 64×64 pixels, coding tree depths of up to four levels (i.e. minimum CU size is 8×8) and TUs varying from 4×4 to 32×32 pixels.

Table 4.1 shows all test cases corresponding to 17 encoding configurations. The baseline encoder configuration is defined as *TEST 1*, while the other 16 configurations correspond to *TEST 2* through *TEST 17*. The table lists the parameter values used in active coding tools, while *D* and *E* represent a disabled or enabled tool/mode, respectively. Although a larger number of tests were performed using a larger spectrum of configuration parameters, only the 16 most representative ones in terms of PSNR, bit rate and computational complexity were selected and included in Table 4.1. Every test was performed using five different QPs: 22, 27, 32, 37 and 42.

Figure 4.1 shows complexity results obtained by encoding the 12 video sequences using the 17 test encoding configurations listed in Table 4.1. In Fig. 4.1, the computational complexity values were normalised with respect to *TEST 1* (reference configuration). For all cases, except *TEST 3*, one can notice a close similarity between the trends of lines in Fig. 4.1, which means that complexity varies fairly likewise for all video sequences when a specific tool is enabled. *TEST 3*, which evaluates the effect of increasing the ME search range from 64 to 128, shows different complexity values for each sequence most likely due to their very different motion characteristics. It is quite evident that video sequences with large motion activity, such as *RaceHorses1* and *BasketballDrive*, result in larger computational complexities than others with little or slow motion, such as *vidyo1*, *vidyo4* and *Traffic*. Encoding efficiency results, measured in terms of bit rate (also normalised with respect to *TEST 1*) and luminance PSNR (Y-PSNR) variation (using *TEST 1* as reference), are shown in Figs. 4.2 and 4.3, respectively, for the same 12 video sequences and 17 test configurations.

Figures 4.1, 4.2 and 4.3 indicate that some configurations do not influence significantly any of the three performance metrics. For example, choosing configuration *TEST 15* has a very small impact on the computational complexity and on bit rate savings in comparison to *TEST 1*, leading to a slight decrease in Y-PSNR for most video sequences. Based on these results, those coding tools that were found to have the largest impact on performance and complexity were selected as the basis for the second step of the computational complexity analysis process, which is described in the next section.

4.1.3 Relevant Encoding Configurations

Most studies on complexity analysis of video encoders focus on testing each feature independently, by comparing the performance of a baseline configuration against the same baseline configuration with only one tool enabled at a time. However, current video coding algorithms are characterised by high levels of interdependency between coding tools, which means that any additional encoding gain obtained by enabling a particular coding option may be dependent on the enabled/disabled status of other coding tools. Since this is the case of HEVC, the complexity analysis presented in this section is based on a sequence of predefined encoder configurations, where the coding tools are enabled in a cumulative order along such sequence. The sequence of relevant encoding configurations was constructed in two steps, as follows.

Table 4.1 Encoder configurations for identifying relevant parameters

Tool	Test case (*TEST*)																
	1	2	3	4	5	6	7	8	9	10	11	12	13	14	15	16	17
Inter 4×4	D	**E**	D	D	D	D	D	D	D	D	D	D	D	D	D	D	D
Search range	64	64	**128**	64	64	64	64	64	64	64	64	64	64	64	64	64	64
Bi-prediction refinement[a]	4	4	4	**8**	4	4	4	4	4	4	4	4	4	4	4	4	4
Hadamard ME	D	D	D	D	**E**	D	D	D	D	D	D	D	D	D	D	D	D
Fast encoding	E	E	E	E	E	**D**	E	E	E	E	E	E	E	E	E	E	E
Fast merge decision	E	E	E	E	E	E	**D**[b]	E	E	E	E	E	E	E	E	E	E
Deblocking filter	D	D	D	D	D	D	D	**E**	D	D	D	D	D	D	D	D	D
Internal bit depth	8	8	8	8	8	8	8	8	**10**	8	8	8	8	8	8	8	8
Sample adaptive offset	D	D	D	D	D	D	D	D	D	**E**	D	D	D	D	D	D	D
Adaptive loop filter	D	D	D	D	D	D	D	D	D	D	**E**	D	D	D	D	D	D
Linear mode intra prediction	D	D	D	D	D	D	D	D	D	D	D	**E**	D	D	D	D	D
Non-square transforms	D	D	D	D	D	D	D	D	D	D	D	D	**E**	D	D	D	D
Asymmetric motion partitions	D	D	D	D	D	D	D	D	D	D	D	D	D	**E**	D	D	D
Transform skipping	E	E	E	E	E	E	E	E	E	E	E	E	E	E	**D**	**D**	E
Fast transform skipping	E	E	E	E	E	E	E	E	E	E	E	E	E	E	E	**D**	E
PCM mode[b]	D	D	D	D	D	D	D	D	D	D	D	D	D	D	D	D	**E**

[a]Search range increase for ME when bi-prediction is used
[b]Pulse code modulation (PCM) mode allows a PU to be encoded with no prediction, no transform and no entropy coding
The bold cases are those that suffered a change in comparison to the previous encoding configuration shown in the table.

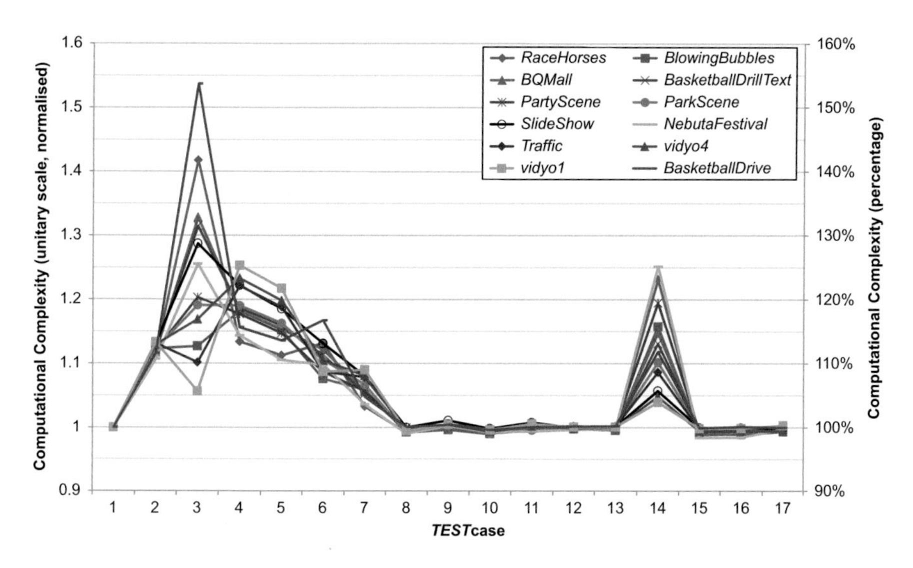

Fig. 4.1 Normalised computational complexity for encoding each video sequence under all 17 configurations (QP 32)

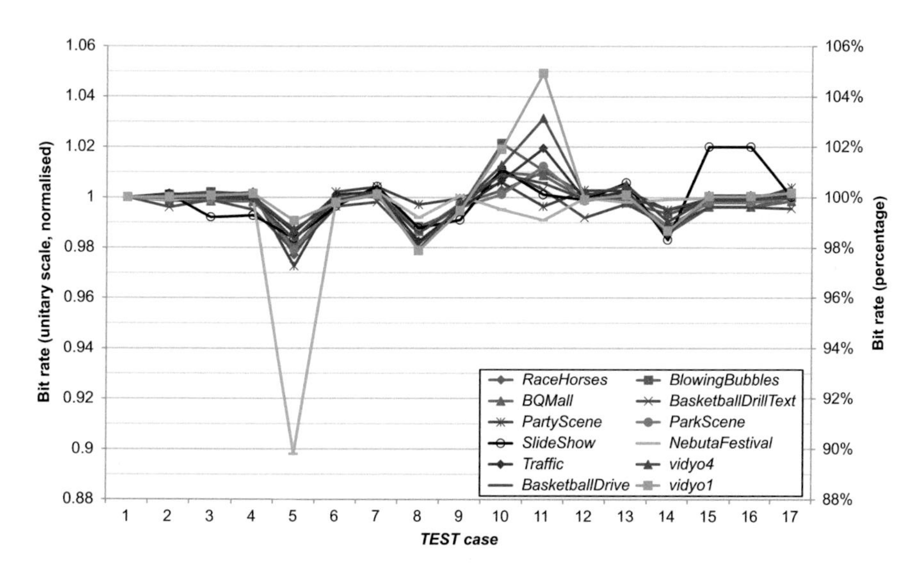

Fig. 4.2 Normalised bit rate for each video sequence encoded under all 17 configurations (QP 32)

In the first step, 13 configurations from Table 4.1 were identified as those having significant impact on Y-PSNR, bit rate and overall computational complexity in comparison to the baseline test case (*TEST 1*). Significant impact was defined here as Y-PSNR variations of at least 0.1 dB, bit rate changes of at least 1.5 % and computational complexity increases of at least 5 %. These 13 configurations correspond to seven coding tools—ME, DBF, SAO, ALF, *internal bit depth* (IBD), *linear mode*

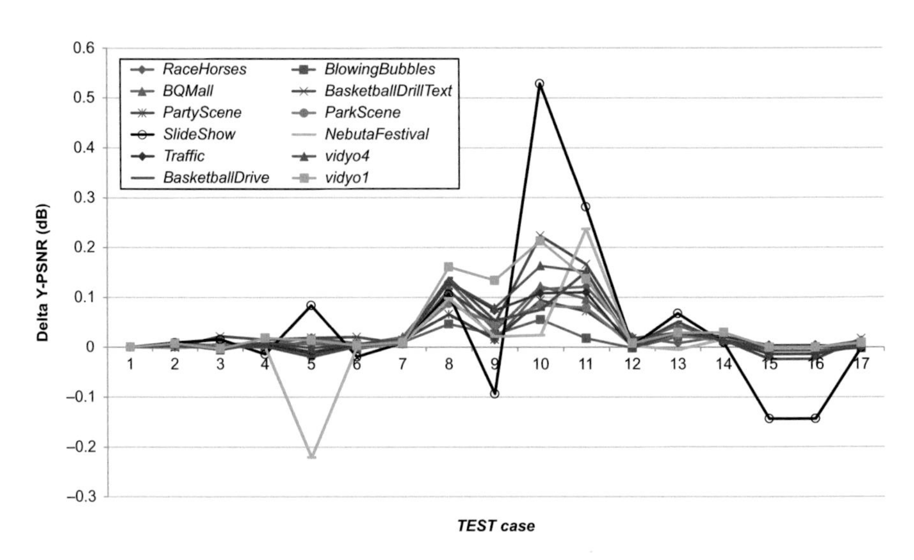

Fig. 4.3 Delta Y-PSNR for each video sequence encoded under all 17 configurations (QP 32)

(LM) *intra-prediction* and *non-square transforms* (NSQT) – which were then used in the second step to create a sequence of 15 relevant configurations.

In the second step, the 15 configurations (*CFG 1* to *CFG 15*) presented in Table 4.2 were created by defining the parameter values for each of the seven tools mentioned above, in a cumulative sequence. An additional configuration (*CFG 16*) was defined by choosing optimal tools and parameters, as shall be explained in detail in the next section. The baseline configuration (*CFG 1*) is the same as *TEST 1* in Table 4.1. Configurations 2, 4, 5, 6, 7, 8, 9 and 10 correspond to different functional modes of ME/MC; configurations 3, 11 and 12 vary filtering operations, enabling the DBF, SAO and ALF, respectively; *CFG 13* increases the IBD from 8 to 10 bits; *CFG 14* enables *LM Intra-Prediction*; and *CFG 15* enables NSQT. As the tools were enabled in a cumulative order, *CFG 1* is the simplest encoder configuration and *CFG 15* is the most complex one, with all tools enabled.

The enabling order of the tools, presented in Table 4.2, was defined based on the ratio between the bit rate reduction and the increase in encoding computational complexity associated with each tool. The ratio between the Y-PSNR decrease and the complexity increase was also considered as a tiebreak whenever the bit rate-complexity ratio was too similar between two tools. Those tools which presented the highest relative coding efficiency-complexity gains were enabled before the others, since they should have higher priority of activation in complexity-constrained video coding systems.

The set of 16 configurations presented in Table 4.2 was used to encode all the test sequences mentioned in Sect. 4.1.1. For each simulation, bit rate, Y-PSNR and complexity results were recorded for use in the performance and complexity trade-off analysis presented in the next section, which evaluates the activation of each HEVC tool/functional mode.

Table 4.2 Encoder configurations used for complexity and performance analysis

Tool	Configuration case (*CFG*)															
	1	2	3	4	5	6	7	8	9	10	11	12	13	14	15	16
Hadamard ME	D	**E**	E	E	E	E	E	E	E	E	E	E	E	E	E	**E**
Deblocking filter	D	D	**E**	E	E	E	E	E	E	E	E	E	E	E	E	**E**
Asymmetric motion partitions	D	D	D	**E**	E	E	E	E	E	E	E	E	E	E	E	**E**
Search range	64	64	64	64	**96**	**128**	128	128	128	128	128	128	128	128	128	**64**
Bi-prediction refinement	4	4	4	4	4	4	**8**	8	8	8	8	8	8	8	8	**4**
Inter 4×4	D	D	D	D	D	D	D	**E**	E	E	E	E	E	E	E	**D**
Fast encoding	E	E	E	E	E	E	E	E	**D**	D	D	D	D	D	D	**E**
Fast merge decision	E	E	E	E	E	E	E	E	E	**D**	D	D	D	D	D	**E**
Sample adaptive offset	D	D	D	D	D	D	D	D	D	D	**E**	E	E	E	E	**E**
Adaptive loop filter	D	D	D	D	D	D	D	D	D	D	D	**E**	E	E	E	**D**
Internal bit depth	8	8	8	8	8	8	8	8	8	8	8	8	**10**	10	10	**8**
Linear mode intra prediction	D	D	D	D	D	D	D	D	D	D	D	D	D	**E**	E	**D**
Non-square transforms	D	D	D	D	D	D	D	D	D	D	D	D	D	D	**E**	**D**

D and E represent a disabled or enabled tool/mode, respectively
The bold cases are those that suffered a change in comparison to the previous encoding configuration shown in the table.

4.1.4 Encoding Performance and Complexity Trade-Off Analysis

This section presents the performance results in terms of Y-PSNR, bit rate and computational complexity for the 16 test cases listed in Table 4.2. The results are summarised in Figs. 4.4, 4.5, 4.6, 4.7 and 4.8 and in Table 4.3.

Computational complexity results for all video sequences under all encoding configurations are plotted in Fig. 4.4, normalised with respect to those of *CFG 1*, as done in Fig. 4.1. In Fig. 4.4 all video sequences exhibit a similar monotonic increase of the encoding complexity from *CFG 1* to *CFG 15*. However, from *CFG 10* to *CFG 15*, the slope is very small, and the normalised computational complexity is approximately constant. The largest computational complexity increases are observed in the transitions from *CFG 4* to *CFG 9*. As shown in Table 4.2, these configurations correspond to different functional modes of ME according to the specified parameters. The results in Fig. 4.4 show that most of the computational complexity increases occur when the ME process is refined.

It is also noticeable in Fig. 4.4 that from *CFG 5* and especially from *CFG 6* onwards, the curves pertaining to the different video sequences start to be farther apart from each other. This happens due to the fact that these two configurations increase the ME search range from 64 to 96 and from 96 to 128, respectively. As explained in Sect. 4.1.3 in regard to *TEST 3* in Fig. 4.1, the video sequences are quite distinct in terms of motion activity, and for this reason they result in different encoder behaviours when the SR used in the ME is increased. The spreading of the curves is much clearer in the case of the configurations with rank order above *CFG 6* due to the cumulative effect of the tool activations. This is observed in the

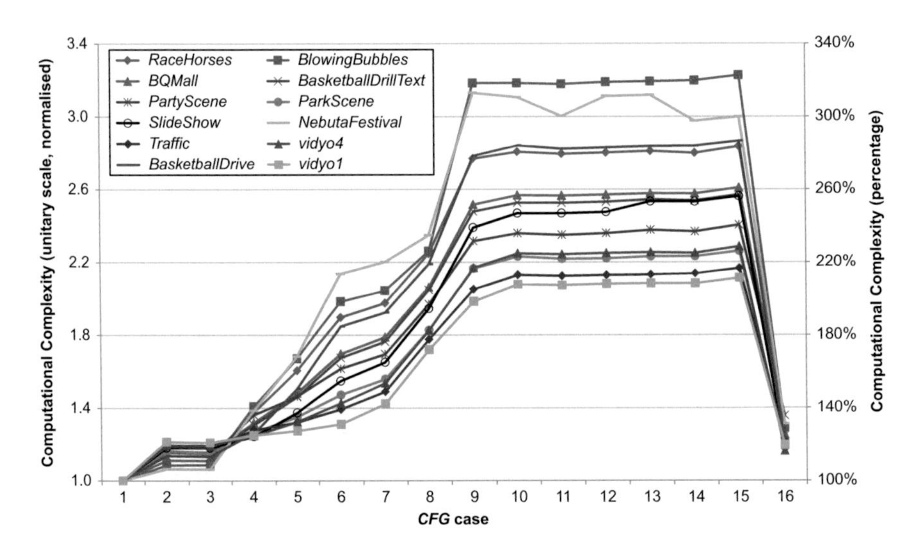

Fig. 4.4 Normalised computational complexity for encoding each video sequence under all 16 configurations (QP 32)

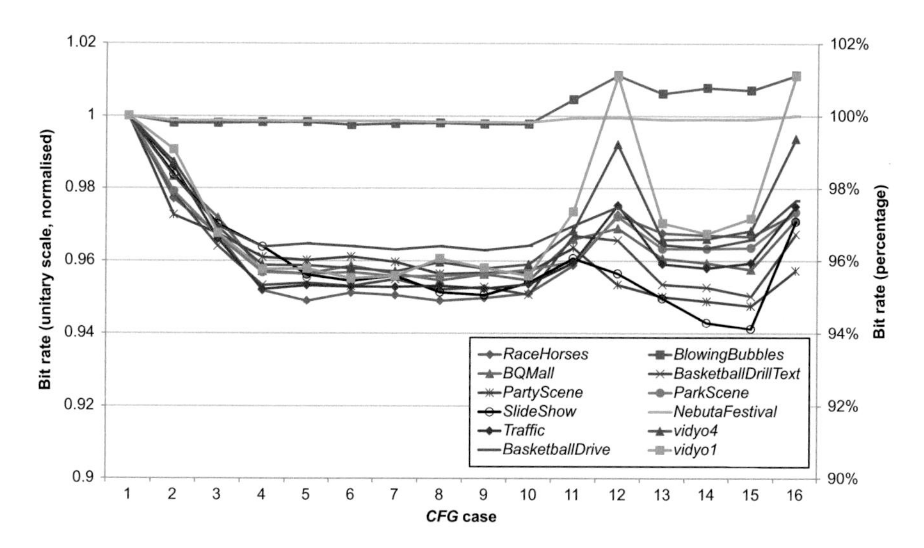

Fig. 4.5 Normalised bit rate for each video sequence encoded under all 16 configurations (QP 32)

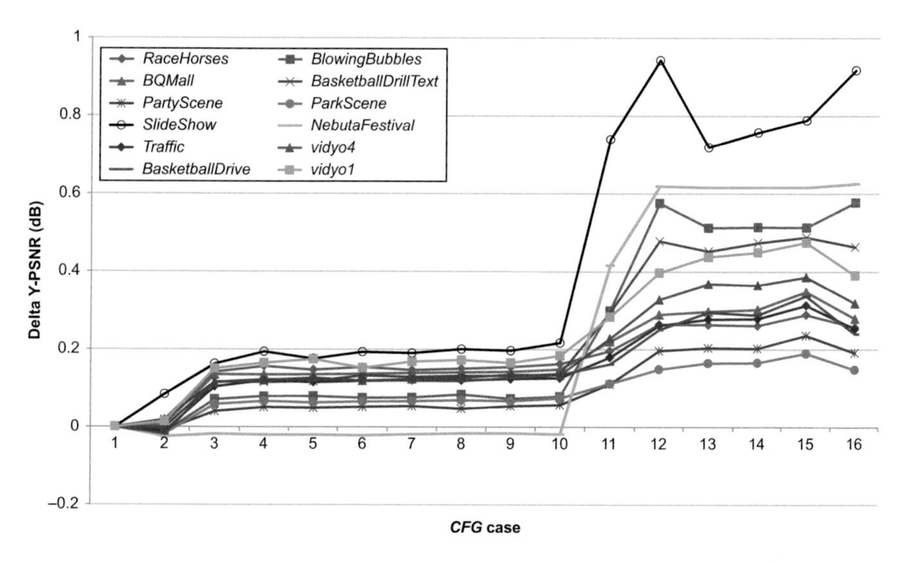

Fig. 4.6 Delta Y-PSNR for each video sequence encoded under all 16 configurations (QP 32)

last configuration, *CFG 15*, for which the computational complexity is up to 3.2 times larger than that of the baseline configuration. Even though Fig. 4.4 presents results for QP 32, other QP values were also tested and showed similar behaviour.

Table 4.3 presents absolute computational complexity values for the least and most complex configurations (*CFG 1* and *CFG 15*, respectively) encoded with QPs 22, 27, 32, 37 and 42. As a reference for comparison, the complexity obtained by encoding the same video sequences with an H.264/AVC high profile encoder

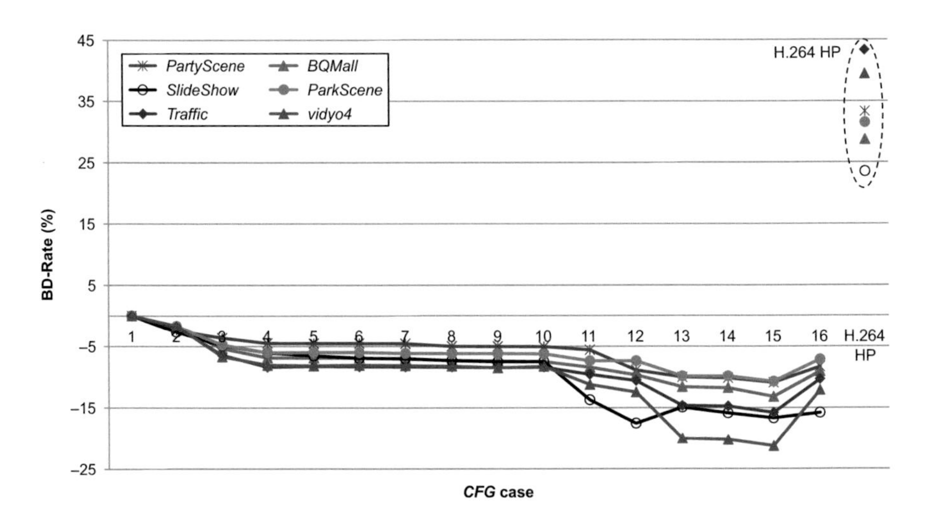

Fig. 4.7 BD-rate values for each configuration using *CFG 1* as reference

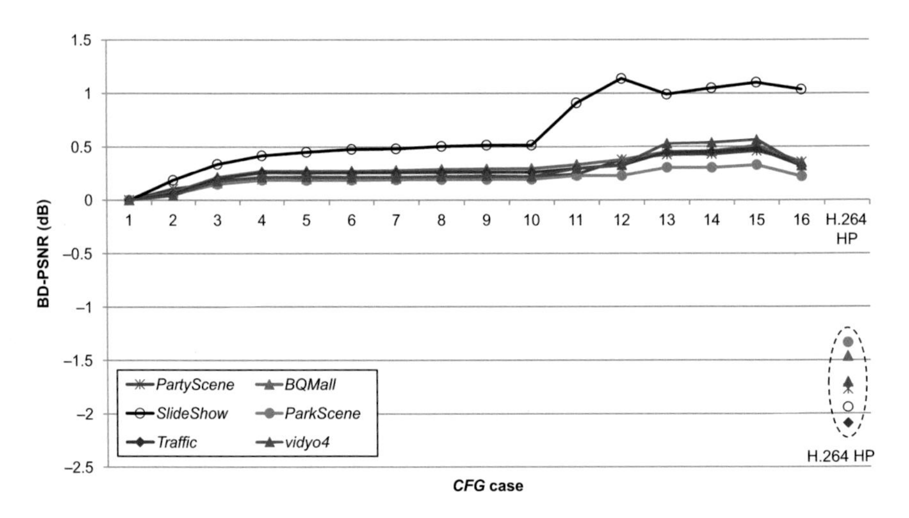

Fig. 4.8 BD-PSNR values for each encoding configuration using *CFG 1* as reference

(JM software, version 18.3) is also presented ("H.264 HP" lines). The table presents results for the sequences *RaceHorses1*, *BasketballDrillText*, *SlideShow*, *ParkScene* and *NebutaFestival*. The rightmost column shows the average complexity increase of *CFG 1* and *CFG 15* over the H.264/AVC HP encoder. It is noteworthy that even the least complex configuration (*CFG 1*) in HEVC is still more complex than H.264/AVC HP (in a range from 9.1 % up to 103.6 %). Moreover, *CFG 15* is at least 1.6 times more complex than H.264/AVC HP, reaching an average computational complexity increase of up to 502.2 % for *NebutaFestival*, which is the most complex

Table 4.3 Computational complexities (in seconds) for HEVC and H.264/AVC configurations under QPs 22, 27, 32, 37 and 42

Video (resolution)	Encoder (Conf.)	QP 22	QP 27	QP 32	QP 37	QP 42	Average Incr. (%)
RaceHorses1 (416×240)	HEVC (CFG 1)	839	740	645	576	1306	18.6
	HEVC (CFG 15)	2293	1969	1828	1599	447	224.5
	H.264 HP	693	603	534	484	1628	–
BasketballDrill Text (832×480)	HEVC (CFG 1)	2493	2226	2023	1870	4085	9.1
	HEVC (CFG 15)	6366	5768	5197	4724	1656	178.4
	H.264 HP	2197	1904	1845	1742	3541	–
SlideShow (1280×720)	HEVC (CFG 1)	4264	4088	3940	3823	8663	46.7
	HEVC (CFG 15)	11,551	10,657	10,093	9581	2567	276.7
	H.264 HP	2857	2718	2639	2610	8684	–
ParkScene (1920×1080)	HEVC (CFG 1)	12,778	11,199	10,299	9676	19,134	15.1
	HEVC (CFG 15)	29,250	25,798	23,283	21,463	9293	160
	H.264 HP	10,310	8743	8385	9040	38,295	–
NebutaFestival (2560×1600)	HEVC (CFG 1)	59,418	57,811	18,065	49,361	111,126	103.6
	HEVC (CFG 15)	190,443	182,055	39,128	142,355	19,154	502.2
	H.264 HP	27,332	24,147	21,132	17,903	1306	–

video sequence among those listed in the table (high spatial resolution, colourful detailed texture and continuous motion activity).

Figure 4.6 shows the increase of Y-PSNR for each encoding configuration in comparison to the reference case. It is possible to notice that the image quality is roughly maintained from *CFG 4* to *CFG 10* and from *CFG 13* to *CFG 15*. Most of the image quality gains are obtained in *CFG 3*, *CFG 11* and *CFG 12*, which enable DBF, SAO and ALF, respectively. These results lead to the conclusion that the three filters have significant impact on the objective image quality in HEVC. When the three filters are enabled (*CFG 12*), the Y-PSNR is increased by up to 0.94 dB (*SlideShow* sequence). Nevertheless, despite the impact on the image quality, it is important to remark that the activation of ALF resulted in a very large bit rate increase, as presented in Fig. 4.5 (see *CFG 12*).

Further results that consider a wider range of QP values (22, 27, 32 and 37) are presented in Figs. 4.7 and 4.8 in terms of BD-rate and BD-PSNR, respectively. Six video sequences were used in the tests, and the results were computed by taking

CFG 1 as the reference case. Both figures reveal large differences between *H.264/AVC HP* (non-connected points on the right side of the figures) and the remaining HM encoding configurations in terms of encoding efficiency. All HM configurations result in a significant decrease in the BD-rate value when more encoding tools are activated, as shown in Fig. 4.7. On the contrary, the *H.264/AVC HP* configuration presents BD-rate increases that vary from 23.5 % to 43.3 %, depending on the video sequence, in comparison to HM *CFG 1*. Similarly, BD-PSNR in Fig. 4.8 is improved when more tools are added to the baseline HM configuration. On the other hand, the *H.264/AVC HP* configuration presents BD-PSNR degradations in a range from 1.3 dB to 2.1 dB in comparison to *CFG 1*.

A relevant conclusion of this study is that the Hadamard ME, the AMP and the filters (except for ALF, which causes a large bit rate increase) should be first enabled in a complexity-constrained encoder, since they increase significantly the encoding efficiency at the cost of a low complexity increase, as previously shown. In fact, most of the gains in the BD-PSNR and BD-rate curves (Figs. 4.7 and 4.8) are accounted for by the tools enabled in *CFG 1-CFG 3* and from *CFG 11* onwards.

The conclusions of this study should be taken into account when selecting the tools and functional modes to define efficient HEVC encoder configurations. The presented results show that high encoding performance under low computational complexity levels can be achieved if the tools and functional modes are carefully chosen.

To exemplify how the presented analysis can be used for tuning an HEVC encoder, an optimised configuration was created. This configuration, which appears in Table 4.2 labelled as *CFG 16*, does not use any tool or functional mode that is not advantageous from the point of view of complexity versus performance. More specifically, in *CFG 16* only the three first parameters listed in Table 4.2 are enabled, which are those that presented the highest relative coding efficiency-complexity gains in the configurations of Table 4.1 (i.e. the ratio between the bit rate reduction and the increase in encoding computational). These parameters are the Hadamard ME, the DBF and the AMP, respectively. Besides them, the SAO filter was also enabled in *CFG 16* because, although it was not classified at the top of the ranking, it yielded significant increases in terms of image quality (Y-PSNR), as shown in Figs. 4.6 and 4.8 (*CFG 11*). The remaining ME functional modes, the IBDI, the *LM Intra-Prediction* and the use of NSQT were all either disabled or enabled in their lowest complexity functional mode, as shown in the last column of Table 4.2.

The results in Figs. 4.4, 4.5, 4.6, 4.7 and 4.8 show that, even though the encoding efficiency of *CFG 16* is similar to that achieved by high-complexity configurations, its computational complexity is much smaller. For instance, *CFG 16* provides roughly the same coding efficiency as *CFG 12*, but its complexity is up to 2.5 times smaller (see Fig. 4.4). This optimisation exercise shows that a wise selection of coding parameters can be used to define a low complexity configuration which is capable of achieving roughly the same coding efficiency as a more complex one. In other words, higher levels of compression efficiency are not necessarily obtained using the most complex coding configurations.

4.1.5 Performance and Complexity as a Function of QP

This section presents an analysis of the encoding performance and computational complexity for different bit rates. In order to decouple the results from the effects of rate control algorithms, a set of fixed QP values (22, 27, 32, 37) was used in each experiment.

Figure 4.9 shows the R-D encoding efficiency of the HM encoder under different configurations for the *BQMall*, *vidyo4*, *ParkScene* and *Traffic* video sequences. The H.264/AVC HP encoder was also included in this evaluation to provide an anchor for comparison. Even though all HM configurations presented in Sect. 4.1.3 were analysed, only four of them (1, 12, 15 and 16) are presented in the charts of Fig. 4.9 for clarity. The results confirm that the compression efficiency of HM configurations is much higher than that of H.264/AVC HP, since the bit rates are reduced by approximately 50 % while maintaining roughly the same Y-PSNR. As *CFG 1* and *CFG 15* are the configurations which present the worst and the best R-D efficiency results, respectively, the performance curves for the remaining HM configurations fall between those of *CFG 1* and *CFG 15*. For this reason, they are omitted from the charts in Fig. 4.9 for clarity. The curves corresponding to *CFG 12* and the optimised configuration *CFG 16* are overlapped in all charts of Fig. 4.9, confirming that their R-D efficiency is approximately the same for all tested QPs.

Figure 4.10 shows the encoding time as a function of QP for the same sequences and the same configurations presented in Fig. 4.9. Figure 4.10 shows that, although *CFG 16* achieves R-D efficiency close to that of *CFG 12* in Fig. 4.9, its computational complexity is much more similar to that of the baseline configuration (*CFG 1*) for all QPs and all video sequences analysed. In fact, the computational complexity of *CFG 16* is closer to that observed in the H.264/AVC HP encoder than to the observed in *CFG 12*, even though its R-D efficiency is almost as high as in *CFG 12*. Therefore, the previous conclusions can be extended for a wide range of bit rates, meaning that coding complexity and efficiency are not necessarily correlated. Thus, in complexity-constrained encoder implementations, it is worthwhile to take these findings into account.

Concerning the effect of QP on the encoding complexity, it was further observed from Fig. 4.10 that the overall complexity decreases slightly when the QP increases. This effect is more prominent in the cases with larger complexity values, which allow such difference to be more easily noticed (*CFG 12* and *CFG 15*). This effect might result from the smaller amount of processed data in the case of higher QP (e.g. more zero coefficients).

4.1.6 Experiment Conclusions

The trade-off between computational complexity and encoding performance of the HEVC encoder was evaluated in the previous sections using a broad range of encoding configuration cases over a wide variety of video contents. The experimental study analysis shows that maximum HEVC complexity can be reduced at practically

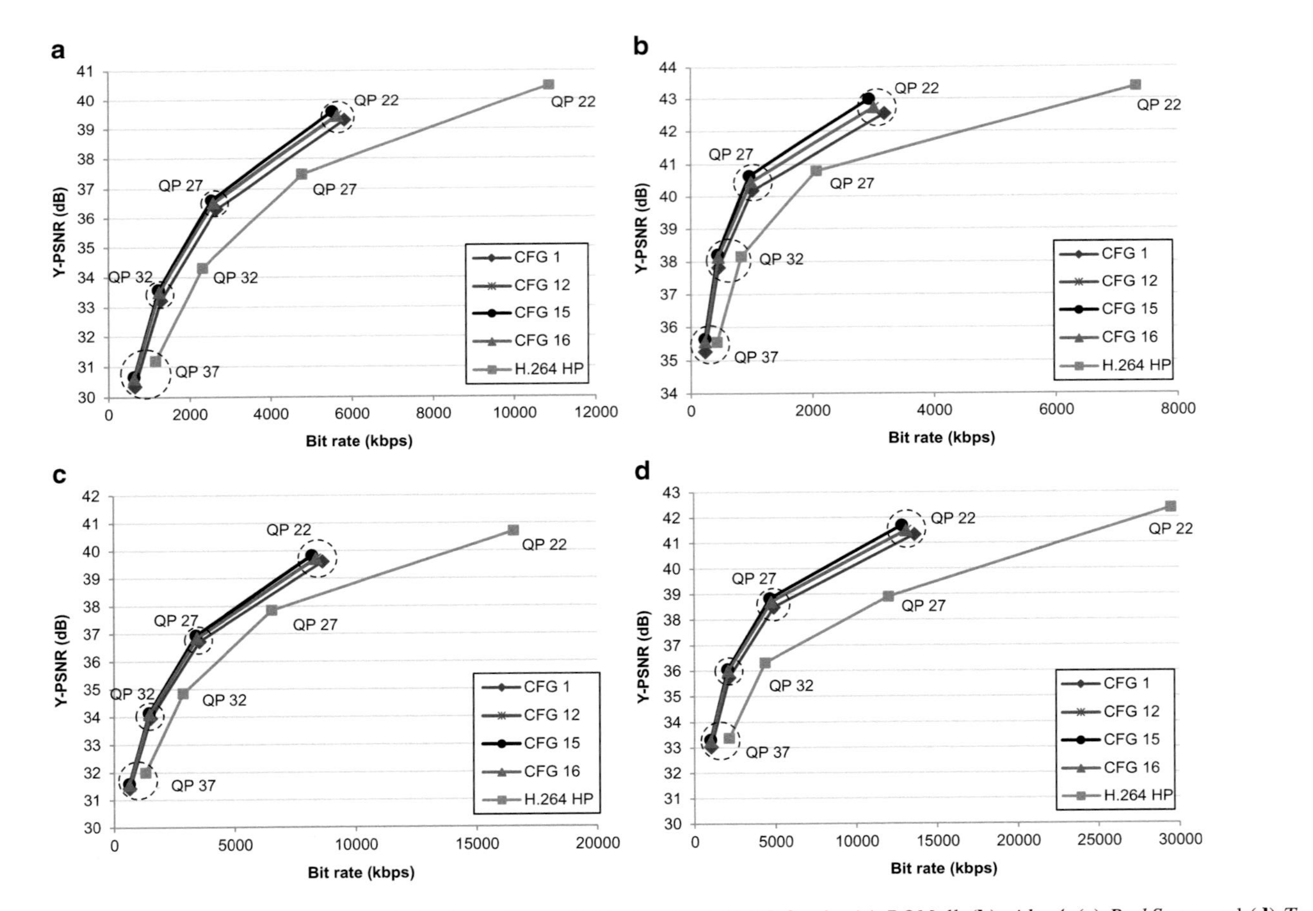

Fig. 4.9 R-D efficiency of HEVC with *CFG 1*, *CFG 12*, *CFG 15* and *CFG 16* and H.264 HP for the **(a)** *BQMall*, **(b)** *vidyo4*, **(c)** *ParkScene* and **(d)** *Traffic* videos (QPs 22, 27, 32, 37)

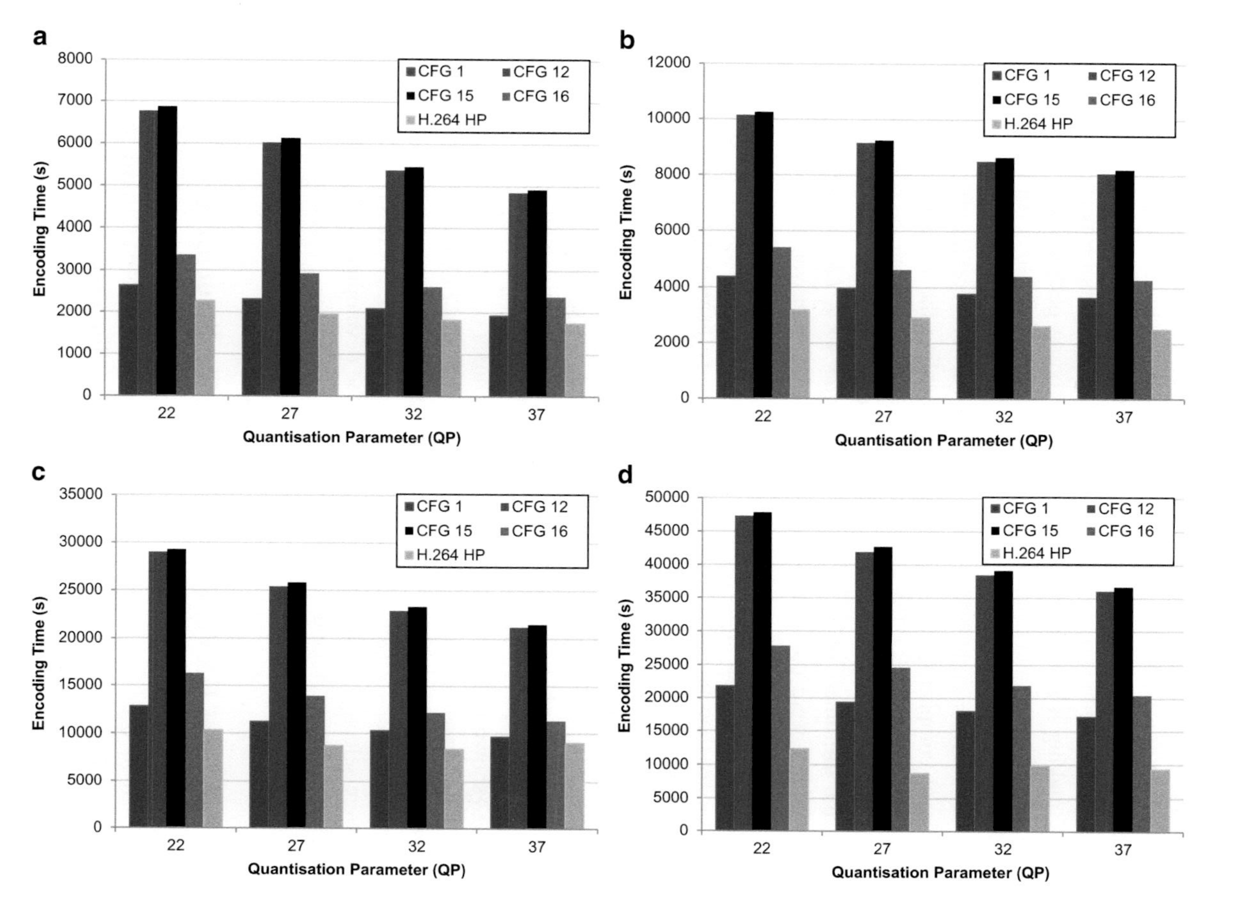

Fig. 4.10 Encoding time of HEVC with *CFG 1*, *CFG 12*, *CFG 15* and *CFG 16* and H.264 HP for the (**a**) *BQMall*, (**b**) *vidyo4*, (**c**) *ParkScene* and (**d**) *Traffic* videos (QPs 22, 27, 32, 37)

no coding efficiency cost, if the coding tools are wisely chosen, combined and configured.

The experimental results show that when the number of tools and functional modes increase in a cumulative progression, the computational complexity grows in a similar way, even though the encoding performance does not increase at the same pace. Therefore, it is advisable to enable first those tools which provide most gains for the least cost. Such a strategy for enabling tools and choosing encoding parameter values lead to a good trade-off between computational complexity and encoding efficiency, making it possible to achieve high encoding performance while still reducing the computational complexity in comparison to the case in which all tools are enabled.

The results demonstrate that a good trade-off between coding efficiency and computational complexity can be achieved by enabling the Hadamard ME, the AMP and the use of filters (DBF and SAO), instead of enhancing the performance of other computationally demanding but not so efficient tools. The conclusions regarding the usefulness of some tools such as *Inter 4×4*, *LM Intra-Prediction*, ALF and NSQT were confirmed later when JCT-VC removed them from HM versions later than 7. The experiments performed with HM7, presented in this chapter, had shown that those tools should be disabled (see the proposed *CFG 16* in Table 4.2), since they did not offer significant encoding efficiency when considering their associated computational complexity increases.

The study presented in this section provided an important basis to devise the R-D-C-optimised control system, presented in Chap. 7 of this book. As it will be explained later on, an R-D-C analysis was performed on a set of encoding configurations, which varied the value of those parameters identified in the previous sections as those that most affect the computational complexity of an HEVC encoder. As shown in Fig. 4.4, the largest computational complexity increases are observed in the transition from *CFG 1* to *CFG 2* and from *CFG 4* to *CFG 9*, which correspond to the *Hadamard ME*, the AMP, the *Search Range*, the *Bi-prediction Refinement*, the *Inter 4×4* and the *Fast Encoding* parameters. For this reason, some of them were selected for the analysis that led to the development of an encoding time control system presented later on in Chap. 7.

4.2 Analysis of HEVC Frame Partitioning Structures

As presented in Chap. 3, complexity-aware solutions for HEVC have focused mainly on the decision process of frame partitioning structures such as CUs, PUs and TUs. However, a systematic analysis on the impact of constraining the decision of such structures in both the R-D efficiency and the encoding computational complexity is still missing in the technical literature. The analysis presented in this section is important for identifying research directions and to lead the studies presented later in this book, which focus mainly on those partitioning structures that yield the largest complexity reductions and the best trade-offs between encoding efficiency and computational complexity when constrained.

4.2.1 Experimental Setup and Methodology

The experiments presented here were performed using a setup similar to that described in Sect. 4.1, using the same 12 test video sequences presented in Sect. 4.1.1 and the HM encoder—version 13 (HM13) [5]. As before, the *Microsoft Visual Studio C++ Compiler* was used to compile the encoder with the *Release* mode and the tests were run on *Intel® Xeon® E5520* (2.27 GHz) processors running the *Windows Server 2008 HPC* operating system. The computational complexity was measured in terms of processing time, reported by the *Intel® VTune™ Amplifier XE* software profiler [3]. The experimental study was carried out by changing the configuration of the frame partitioning structures, one parameter at a time, such that the impact of each one could be independently analysed. Every test was performed using five different QPs: 22, 27, 32, 37 and 42.

4.2.2 Frame Partitioning Configurations Tested

As explained in Sect. 2.6.1.2, the main frame partitioning structures of HEVC are CUs, PUs and TUs, which size and format can vary broadly following a quadtree-structured partitioning scheme. There are three frame partitioning parameters in the HEVC encoder that can control directly these structures, *Max CU Depth*, *Max TU Depth* and *AMP*, which define, respectively, the maximum quadtree depth allowed for a CU in each CTU, the maximum quadtree depth allowed for a TU in each RQT and the possibility of using *asymmetric motion partitions* in the PU format decision.

Starting with a baseline encoder configuration using the *random access* temporal configuration, new configurations were created by modifying the value of each partitioning parameter, one at a time in a non-cumulative way (i.e. a parameter was changed in one configuration and then set to its original value in the following configuration). The resulting image quality, bit rate and encoding computational complexity were recorded for comparison with the reference baseline configuration.

The seven configurations tested are presented in Table 4.4. The baseline encoder configuration is defined as *PAR 1*, while the other six configurations correspond to *PAR 2* through *PAR 7*. In *PAR 1*, the maximum CU depth allowed is 4 (*Max CU Depth*), the maximum TU depth allowed is 3 (*Max TU Depth*) and the use of AMP is enabled. From *PAR 2* to *PAR 4*, the only parameter changed is *Max CU Depth*. In *PAR 5* and *PAR 6*, the effect of changing only the *Max TU Depth* parameter is tested. Finally, in *PAR 7* the activation of AMP is evaluated. The values *D* and *E* represent the disabled and enabled states of AMP, respectively.

All test sequences listed in Sect. 4.1.1 were encoded with the seven configurations presented in Table 4.4. For each simulation, bit rate, Y-PSNR and complexity results were recorded in order to allow the performance and complexity trade-off analysis presented in the next section.

Table 4.4 Frame partitioning structure configurations tested in the experiments

	Frame partitioning configuration (PAR)						
Parameter	1	2	3	4	5	6	7
Max CU Depth	4	**3**	**2**	**1**	4	4	4
Max TU Depth	3	3	3	3	**2**	**1**	3
AMP	E	E	E	E	E	E	**D**

The shaded bold values are those that suffered a change in comparison to the previous encoding configuration shown in the table.

4.2.3 Encoding Performance and Complexity Trade-Off Analysis

This section presents the performance results in terms of PSNR, bit rate and computational complexity for the seven test cases listed in Table 4.4. The results are summarised in Table 4.7 and Figs. 4.11, 4.12, 4.13, 4.14 and 4.15. Even though Figs. 4.11, 4.12 and 4.13 present results for QP 32, other QPs were also tested and have shown similar behaviour. General results for QPs 22, 27, 32 and 37 are shown in Figs. 4.14 and 4.15.

Results in terms of computational complexity for all video sequences encoded under all tested configurations are plotted in Fig. 4.11, normalised with respect to *PAR 1*. In Fig. 4.11, all video sequences exhibit a similar variation in the encoding complexity from *PAR 1* to *PAR 7*. The largest complexity decreases are noticed from *PAR 1* to *PAR 4*, which represent the cases when the value of *Max CU Depth* was modified. In *PAR 5* and *PAR 6*, the computational complexity has also decreased in comparison to *PAR 1*, even though in a much smaller extent. Finally, with *PAR 7* the encoding computational complexity experiences a small decrease varying from 1 % to 19 % in comparison to *PAR 1*. It is also noticeable that, differently from the first six configurations, the curves in *PAR 7* are more distant from each other. This happens due to the fact that this configuration affects directly the ME operation, and as the video sequences are quite distinct in terms of motion activity, they result in different encoder behaviours when AMP is disabled. A similar effect has been previously noticed in Fig. 4.4.

Normalised bit rate results are presented in Fig. 4.12. It is noticeable that *PAR 2*, *PAR 3* and *PAR 4*, which are also those configurations which resulted in the largest decreases in computational complexity, are also responsible for the largest bit rate increases. However, even though the computational complexity is affected in a similar way for all video sequences in such cases, the compression efficiency varies differently from one video to another, depending on its characteristics. This happens because when *Max CU Depth* is decreased, the number of tested coding tree possibilities decreases exponentially in all video sequences, but the effect on the compression efficiency is much more noticeable in low-resolution video sequences, which use smaller CU sizes in the encoding process than in high-resolution video sequences, which are generally encoded with larger CU sizes. This is visible in Fig. 4.12, which shows that the largest bit rate increases appear for the *SlideShow*

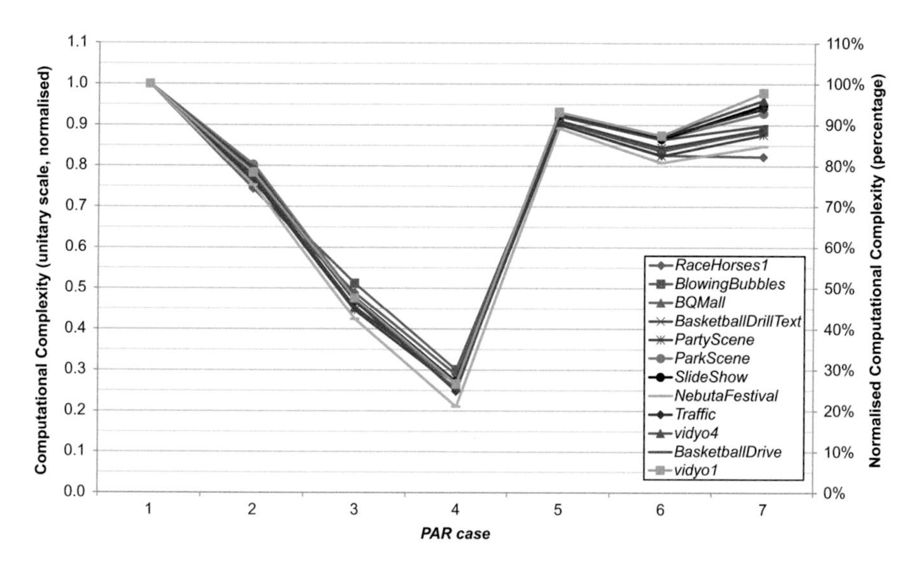

Fig. 4.11 Normalised computational complexity for encoding each video sequence under the seven frame partitioning configurations (QP 32)

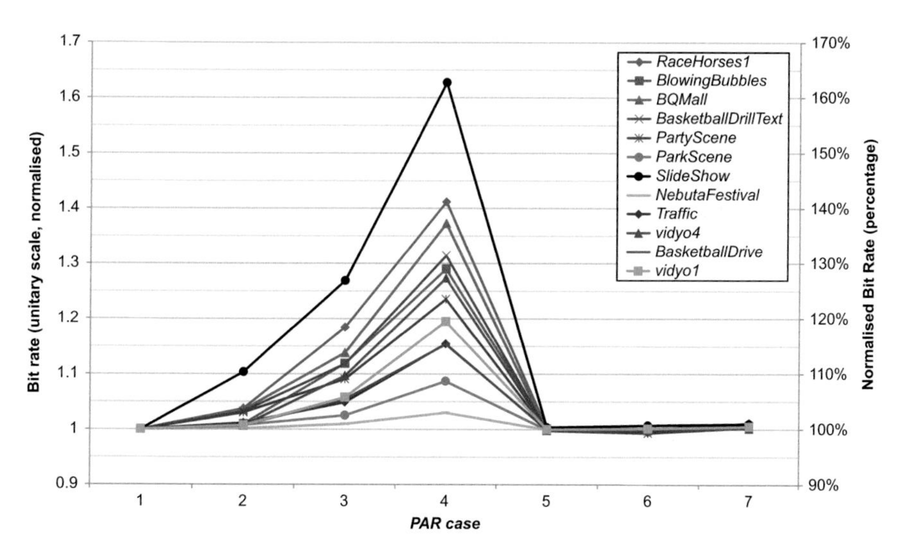

Fig. 4.12 Normalised bit rate for each video sequence encoded under the seven frame partitioning configurations (QP 32)

(1280×720), *RaceHorses1* (416×240) and *BQMall* (832×480) sequences, while the smallest increases appear for *NebutaFestival* (2560×1600), *ParkScene* (1920×1080) and *Traffic* (2560×1600) sequences. The bit rate remains practically the same in *PAR 5*, *PAR 6* and *PAR 7*, in comparison to *PAR 1*.

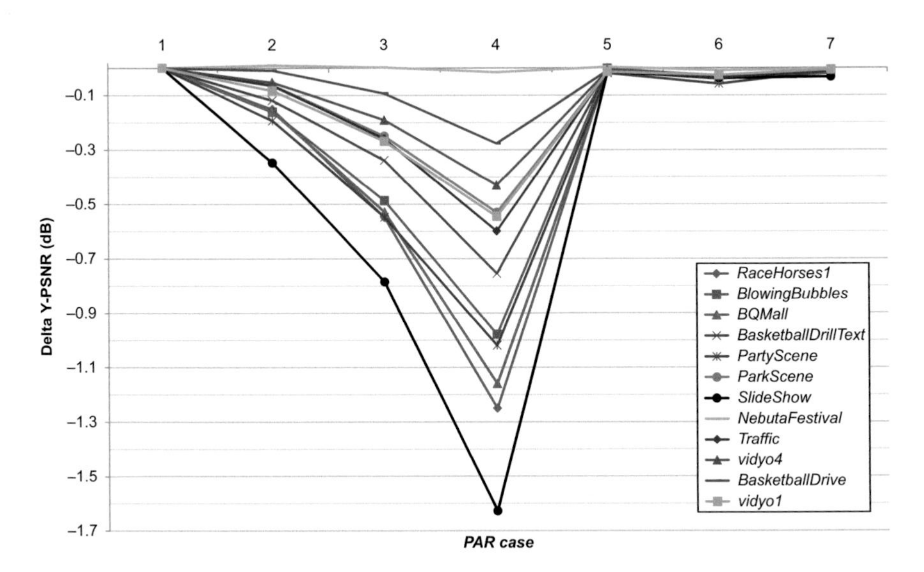

Fig. 4.13 Delta Y-PSNR for each video sequence encoded under the seven frame partitioning configurations (QP 32)

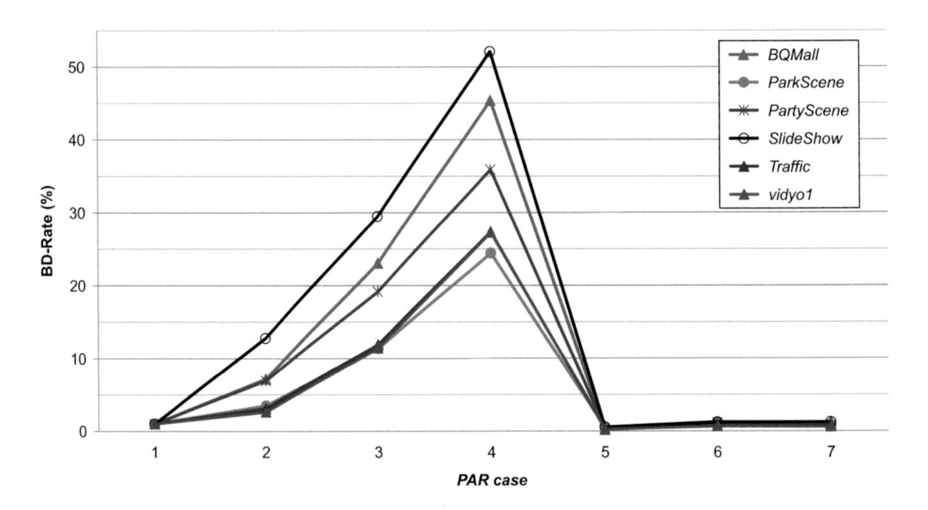

Fig. 4.14 BD-rate values for each configuration using *PAR 1* as reference

Figure 4.13 shows the image quality variation in terms of Y-PSNR for each configuration in comparison to *PAR 1*. It is possible to notice that the quality is maintained almost unchanged from *PAR 5* to *PAR 7*. Most of the image quality drops are noticed in *PAR 2*, *PAR 3* and *PAR 4*, respectively, due to the same reasons explained for Fig. 4.12.

Further results in terms of BD-rate and BD-PSNR are presented in Figs. 4.14 and 4.15, respectively. The BD values were computed using QPs 22, 27, 32 and 37. Six

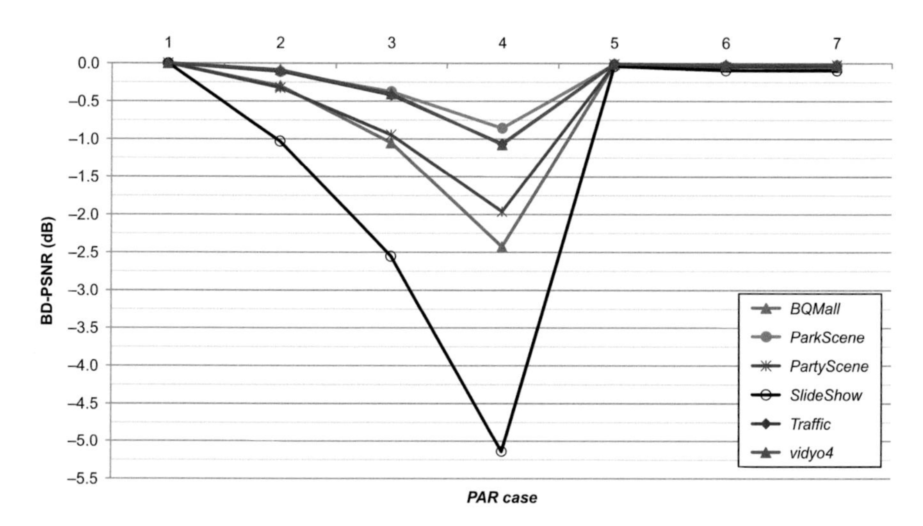

Fig. 4.15 BD-PSNR values for each configuration using *PAR 1* as reference

Table 4.5 BD-rate increase (%) per configuration

	Frame partitioning configuration (PAR)					
Video sequence	2	3	4	5	6	7
PartyScene	6.9	19.1	35.9	0.3	1.1	0.5
BQMall	7.2	23.0	45.3	0.3	1.1	1.2
SlideShow	12.8	29.5	52.1	0.5	1.2	1.2
ParkScene	3.5	11.4	24.4	0.4	1.1	0.8
Traffic	3.1	11.9	27.3	0.2	0.7	0.9
vidyo4	2.2	13.6	30.7	0.2	0.7	0.5
Average	**5.9**	**18.1**	**36.0**	**0.3**	**1.0**	**0.9**

video sequences were used in the tests, and the results were computed by taking *PAR 1* as the reference case. Confirming the results discussed previously for QP 32, the configurations *PAR 2*, *PAR 3* and *PAR 4* are those which reveal the largest decreases in compression efficiency in comparison to *PAR 1*, while *PAR 5*, *PAR 6* and *PAR 7* maintained BD-rate and BD-PSNR results close to zero. In fact, when *Max CU Depth* is set to 1 (*PAR 4*), the BD-rate increases reach values above 24 % and up to 52 % in the worst case, which shows that the indiscriminate decrease of *Max CU Depth* is an extremely inefficient way of reducing the computational complexity of HEVC encoders.

Table 4.5 and Table 4.6 present results in terms of BD-rate increase and average computational complexity decrease per configuration tested, while Table 4.7 shows the ratio between the two values presented in Table 4.5 and Table 4.6. On the one hand, according to Table 4.7, the best trade-off between computational complexity reduction and compression efficiency would be achieved by first using *PAR 5*, *PAR 6* and *PAR 7* and, as last resource, *PAR 2*, *PAR 3* and *PAR 4*. On the other hand,

Table 4.6 Average computational complexity reduction (%) per configuration

	Frame partitioning configuration (PAR)					
Video sequence	2	3	4	5	6	7
PartyScene	23.4	54.7	74.30	9.9	17.6	9.7
BQMall	23.2	55.1	73.4	10.0	17.1	11.8
SlideShow	22.4	52.6	73.1	7.8	13.8	5.2
ParkScene	20.1	52.5	75.0	8.5	15.0	8.5
Traffic	22.1	53.7	75.6	8.3	14.7	7.4
vidyo4	22.2	52.7	73.1	7.6	13.6	5.1
Average	**22.2**	**53.5**	**74.1**	**8.7**	**15.3**	**8.0**

Table 4.7 Ratio between BD-rate increase (%) and computational complexity reduction

	Frame partitioning configuration (PAR)					
Video sequence	2	3	4	5	6	7
PartyScene	29.5	35.0	48.3	2.8	6.3	5.5
BQMall	30.9	41.8	61.7	2.9	6.4	10.2
SlideShow	57.1	56.1	71.3	6.9	8.9	23.7
ParkScene	17.2	21.6	32.6	5.1	7.4	9.7
Traffic	14.0	22.2	36.2	2.4	5.0	12.1
vidyo4	9.8	25.7	42.0	2.2	5.2	9.3
Average	**26.4**	**33.7**	**48.7**	**3.7**	**6.5**	**11.8**

choosing *PAR 5*, *PAR 6* or *PAR 7* would allow a maximum computational complexity decrease of only 15.3 %, which is not sufficient when the large computational complexity of HEVC is taken into account.

4.2.4 Experiment Conclusions

The previous section presented an analysis on the trade-off between computational complexity and encoding performance of HEVC when different constraints are applied to the decision of the frame partitioning structures. Seven configurations and 12 video sequences were used in the tests. The analysis shows that the encoding computational complexity can be thoroughly reduced by managing the frame partitioning structures, but some configurations incur in much larger costs in terms of compression efficiency than others.

By changing the maximum RQT depth allowed and the possibility of using AMP (i.e. *PAR 5*, *PAR 6*, *PAR 7*), the HEVC encoding complexity can be reduced at very small compression efficiency costs. The costs of modifying such structures varied between 3.7 and 11.8, while the costs of decreasing complexity by changing the maximum CU depth (i.e. *PAR 2*, *PAR 3*, *PAR 4*) varied between 26.4 and 48.7, as presented in Table 4.7. However, as previously noticed, the computational complexity decrease achieved when changing the first structures mentioned varied between 8.0 % and 15.3 %, which is still modest when considering the enormous computational complexity of HEVC (up to 502 % more complex than H.264/AVC HP, as shown in Table 4.3).

In order to achieve larger complexity reductions levels minimising the compression efficiency loss, the encoder should be able to choose wisely their parameters according to the video characteristics, so that a positive trade-off between computational complexity and compression efficiency is achieved. Instead of simply removing indiscriminately the possibility of using certain frame partitioning structures in the whole video sequence, as done in the experiments described in the previous sections, the encoder must be able to decide the constrained structure at a finer scale, such as in a per-frame or per-CU frequency, in response to the time- and space-varying characteristics of the video signal.

The next chapters of this book present the authors' contributions for efficiently reducing and scaling the computational complexity of HEVC encoders.

References

1. G. Correa, P. Assuncao, L. Agostini, L.A. da Silva Cruz, Performance and computational complexity assessment of high efficiency video encoders. IEEE Trans. Circ. Syst. Video Technol. **22**, 1899–1909 (2012)
2. ISO/IEC-JCT1/SC29/WG11, High efficiency video coding (HEVC) test model 7 (HM 7) Encoder Description, Geneva, Switzerland (2012)
3. VTune™ Amplifier XE from Intel. Available: http://software.intel.com/en-us/articles/intel-vtune-amplifier-xe/
4. ISO/IEC-JCT1/SC29/WG11, JCT-VC AHG report: complexity assessment (AHG 12), Geneva, Switzerland (2011)
5. ISO/IEC-JCT1/SC29/WG11, High efficiency video coding (HEVC) test model 13 (HM 13) Encoder Description, in Geneva, Switzerland (2013)

Chapter 5
Complexity Scaling for HEVC Using Adaptive Partitioning Structures

This chapter presents a set of methods which were developed with the goal of dynamically scaling the computational complexity required by HEVC in the encoding process [1–8]. All of them rely on constraining the frame partitioning structures introduced by the standard, namely, the CUs and the PUs, in order to adjust the number of R-D evaluations performed in the optimisation process and consequently the encoding computational complexity.[1]

By using the proposed algorithms, the encoding computational complexity can be downscaled by up to 50 % with negligible R-D performance losses and down to 20 % of the unconstrained complexity with larger losses.[2] The levels of performance degradation observed when applying these complexity scaling methods are acceptable in many applications and in power-constrained devices where some sort of encoding complexity reduction methods have to be applied.

5.1 Introduction and Motivation

The study carried out in Chap. 4 has shown that the computational complexity of HEVC is a direct consequence of the frame partitioning structures that lead to nested encoding loops, such that the encoding of CUs at deeper coding tree depths is a process invoked during the encoding of CUs at lower depths. For each CU in each possible coding tree configuration, all possible PU divisions and all possible

[1] The experiments presented in this chapter were performed using HM versions from 2 to 9. However, since the results for each proposed algorithm are compared against the unmodified HEVC encoder of the same version, the differences between encoder versions can be disregarded. Besides, the frame partitioning structures did not suffer significant modifications during the standardisation process.

[2] As in Chap. 4, computational complexity was measured in terms of processing time by the *Intel® VTune™ Amplifier XE* software profiler.

G. Corrêa et al., *Complexity-Aware High Efficiency Video Coding*,
DOI 10.1007/978-3-319-25778-5_5

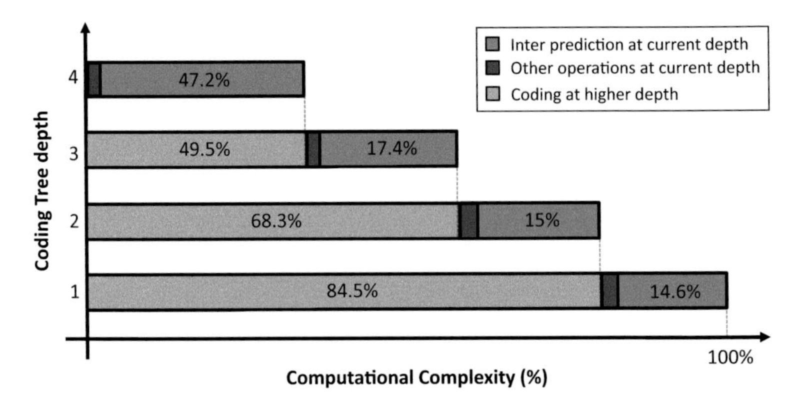

Fig. 5.1 Computational complexity of encoding CUs in each coding tree depth

RQT formations are tested in an RDO process, which considers every encoding possibility and compares all of them in terms of R-D efficiency. As the R-D cost for each encoding possibility is computed only after intra-/inter-prediction, direct and inverse transform and quantisation, entropy coding and filtering operations, the complexity of defining the best combination of CU, PU and TU structures is the bottleneck of the HEVC encoder.

Chapter 4 has also shown that CUs are the most suitable frame partitioning structures to be adjusted when large amounts of computational complexity reductions are sought, even though with potentially large costs in terms of compression efficiency reduction. Figure 5.1 shows the distribution of average computational complexity of CU encoding as a function of tree depth for HM. Coding tree depths are presented in the *y*-axis, and the computational complexity is presented in the *x*-axis. The figure reflects explicitly the nested nature of the HEVC encoding process, with CUs at higher tree depths encoded inside CUs at smaller tree depths. For each CU depth (*y*-axis), the computational complexity is divided into three components: (a) the complexity of performing inter-frame prediction for CUs at the current depth (in blue), (b) the complexity of coding the same image area as CUs at a higher depth (in grey) and (c) the complexity of other operations at the current depth (in red).

Due to this nested encoding structure, the percentage of computational complexity for encoding CUs at the first coding tree depth is indicated as 100 % in Fig. 5.1. It is divided into the complexity of coding CUs at the second tree depth (84.5 %, in grey), the complexity of inter-prediction for 64×64 CUs (14.6 %, in blue) and the complexity of other small operations (0.9 %, in red). Similarly, the complexity of encoding CUs in the second tree depth includes the complexity of encoding CUs at the third tree depth (in grey), the complexity of inter-prediction for 32×32 CUs (in blue) and the complexity of other various operations (in red). The same structure applies to the third depth. Finally, for coding tree depth = 4, the complexity includes only the inter-prediction for 8×8 CUs (in blue) and the complexity associated with other small operations (in red), since no further tree depths are allowed.

Notice that although the smallest CU depths show the largest total percentages of computational complexity in Fig. 5.1, the highest CU depths are the actual responsible for most of the encoding computational complexity (e.g. 49.5 % when coding tree depth=4). For example, although compressing CUs in the third depth (composed by 16×16 CUs) represents 68.3 % of the overall computational complexity, in fact only 18.8 % is the real complexity associated to this CU size, whereas almost all the remaining complexity (49.5 %) is dedicated to operations at the fourth depth.

The behaviour of CU coding along the temporal domain was also investigated. Based on the concept of temporal stationarity (also called *stillness*), defined as the tendency of a video sequence to comprise large image areas with either no or low motion between frames, an experimental study was carried out in order to characterise coding tree depth variations in co-localised areas of neighbouring frames. Figure 5.2 represents the maximum coding tree depth variation used in random co-localised 64×64 areas for 50 frames of the (a) *BQTerrace*, (b) *BasketballDrive* and (c) *Cactus* video sequences, which were encoded with QPs 27, 32 and 37. The figure shows that the maximum coding tree depth does not change very often in the video sequences, which means that once a depth is used in a determined area of the video, the same depth tends to be used in co-localised areas of temporally adjacent frames before it changes to a different value. Based on this observation, the first methods proposed in this chapter were developed assuming that maintaining the maximum coding tree depth for a relatively long period and skipping all the R-D tests at deeper tree levels would incur in a small decrease in terms of R-D performance.

5.2 Fixed Depth Complexity Scaling

The fixed depth complexity scaling (FDCS) method [1] is proposed here as the simplest way of adjusting the maximum coding tree depth used in a frame according to the system's computational limitations. Initially, the encoding process is performed normally using the maximum coding tree depth possible for all CTUs in a frame. Then, whenever the computational complexity increases beyond an upper bound, the maximum depth allowed for CTUs in the next frame is decremented by one unit. Oppositely, when the complexity decreases to a value under the upper bound, the maximum depth allowed is incremented by one unit. The process is repeated if necessary until the minimum or the maximum depth allowed by the standard is reached or the target computational complexity is achieved.

5.2.1 Algorithm Overview

In this section, the method is explained in more detail with the help of the high-level diagram presented in Fig. 5.3. Initially, the video sequence is partitioned into temporal video segments composed of N consecutive frames. Then, the first three frames at the beginning of each video segment are encoded using the maximum

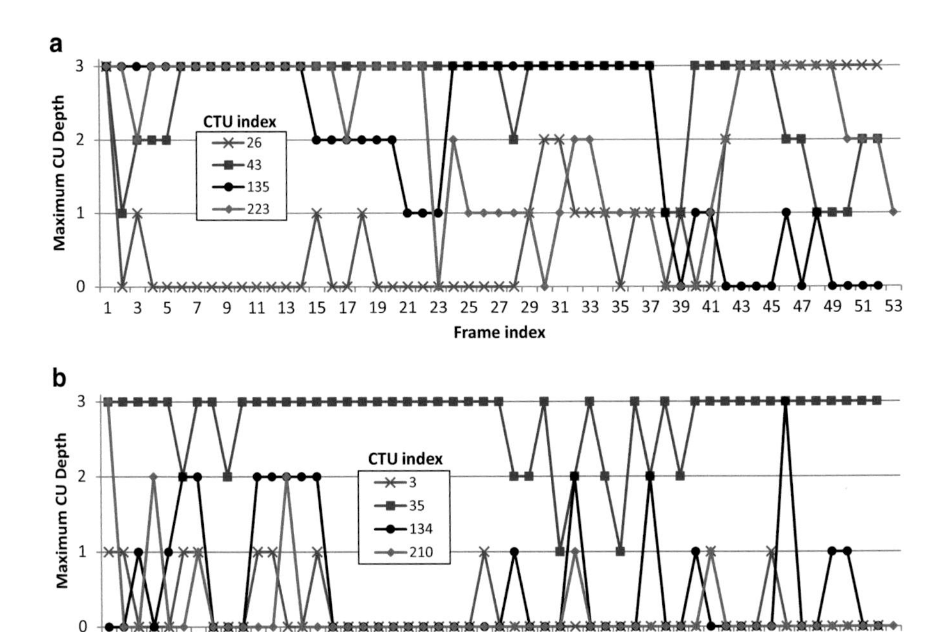

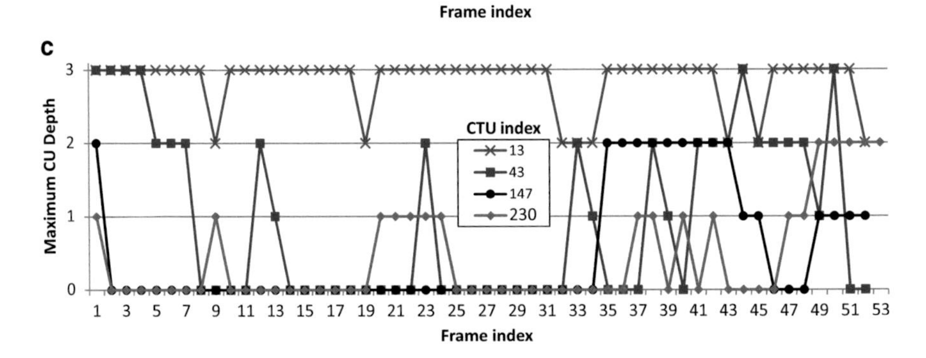

Fig. 5.2 Maximum coding tree depth of random co-localised 64 × 64 areas for sequences (**a**) *BQTerrace* (QP 27), (**b**) *BasketballDrive* (QP 32) and (**c**) *Cactus* (QP 37)

possible coding tree depth. These are called unconstrained frames (Fu). The time spent to encode each Fu is used to estimate the maximum overall computational complexity for the video segment, as shown in Eq. (5.1), where CE^{Max} is the estimated maximum complexity, CS^{Fi} is the computational complexity spent to encode the *i*th frame and *N* is the number of frames in the video segment to be encoded. The average complexity of the first three frames was used in the calculation of CE^{Max} in order to take into account possible computational complexity variations caused by the use of a different number of reference frames in each image in a GOP.

$$CE^{Max} = \frac{N}{3}\sum_{i=1}^{3} CS^{Fi} \tag{5.1}$$

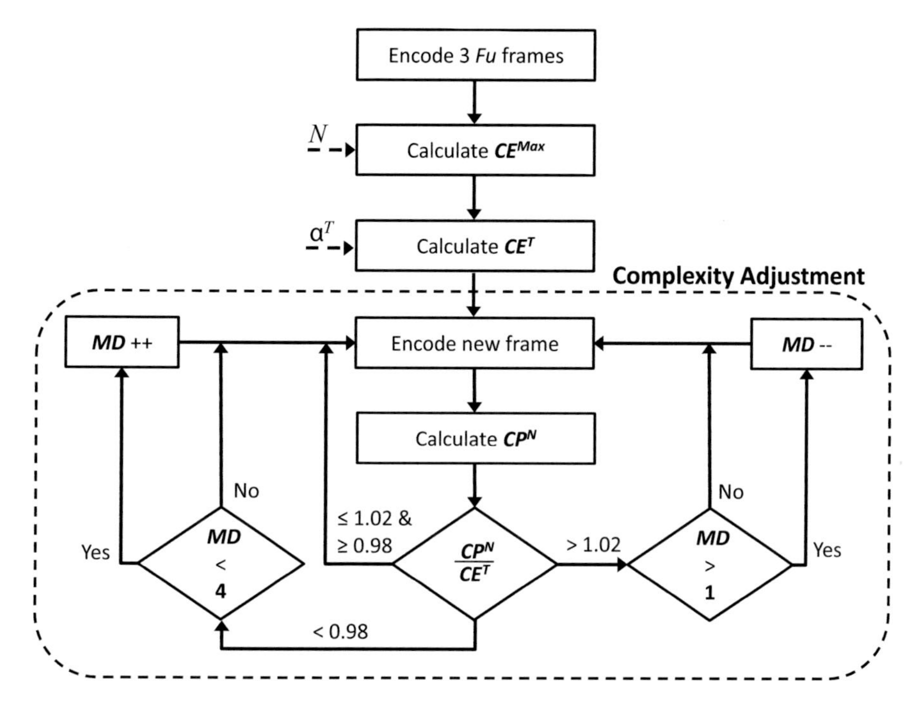

Fig. 5.3 Diagram for the FDCS algorithm

The value of CE^{Max} is then used to set the maximum target complexity CE^{T} available to encode the whole video segment, as defined in Eq. (5.2), where $\alpha^{T} \in [0\ \%, 100\ \%]$ is the complexity reduction ratio defined either as a user parameter or computed from system parameters (e.g. remaining battery life). In this case, 100 % represents the maximum possible encoding complexity when no reduction is imposed.

$$CE^{T} = \alpha^{T} \cdot CE^{Max} \tag{5.2}$$

After defining the target complexity CE^{T}, the algorithm keeps encoding Fu frames and maintains a record of the computational expenditure while encoding the video segment. The record is used to estimate the complexity for the whole segment (CP^{N}), which is calculated after encoding each frame. The CP^{N} calculation assumes that the complexity of encoding the next frames in the segment is similar to the complexity of the most recently encoded frame. CP^{N} is computed as shown in Eq. (5.3), where CS^{Fi} is the computational complexity spent to encode the ith frame of the segment, CS^{FNE} is the complexity spent to encode the last frame and NE is the number of frames already encoded in the video segment.

$$CP^{N} = \sum_{i=1}^{NE} CS^{Fi} + CS^{F_{NE}} \cdot (N - NE) \tag{5.3}$$

While encoding the video segment, if CP^N is lower than the limit imposed by CE^T, the next frame is encoded as Fu. However, when CP^N increases beyond the limit imposed by CE^T, the maximum depth allowed in the next frame (MD, in Fig. 5.3) is decremented by one unit (if the minimum depth has not been achieved yet). In such case, the frame is called a constrained frame (Fc). When CP^N decreases to a value under the limit imposed by CE^T, the maximum depth allowed in the next frame is incremented by one unit with possible saturation at the maximum depth allowable by the standard. The maximum depth value is maintained if CP^N is equal (or very close) to CE^T. To avoid adjustments caused by small computational complexity differences between adjacent frames (which occur naturally in video sequences due to their intrinsic changing characteristics), a difference of 2 % is accepted between CP^N and CE^T without triggering a change in the MD value, as shown in Fig. 5.3. This value was defined after experimental observations.

5.2.2 *Results for FDCS and Performance Evaluation*

The FDCS method was evaluated by measuring the encoding computational complexity reduction accuracy under specific targets and its influence on the R-D performance of the HM encoder (HM4). Three video sequences comprising 500 frames (*BasketballDrive*, *BQTerrace*, *Cactus*) and a concatenation of two of them (*BasketballDrive* and *Cactus*, from now on referenced as *BasketballCactus*) were used in the experiments. The *low delay P* temporal configuration was used in all tests. The encoder performance was evaluated under five complexities reduction ratios (from 60 % to 100 % with 10 % steps) and four different QP values (27, 32, 37 and 42).

The R-D efficiency as a function of the target computational complexity is shown in Fig. 5.4. Average values for bit rate (kbps) and luminance PSNR (dB) for the four test sequences are presented. Table 5.1 shows the average performance results in terms of running complexity, BD-rate and BD-PSNR for the five target complexities tested. The relative bit rate increases, and the Y-PSNR reductions are calculated with reference to the maximum complexity (100 %) results, which correspond to the case in which no complexity scaling is applied.

As expected, the bit rate increases when small target complexities are used. In the worst case (α^T set to 60 %), a BD-rate increase of 12.02 % was noticed. This happens due to the fact that limiting the coding tree depth to decrease the computational complexity leads to a smaller number of small-sized CUs, which results in prediction residues with more spatial structure and larger magnitudes, which are more difficult to encode, leading to higher bit rates. As small blocks are not allowed in such cases, a coarser prediction is performed, which also results in a lower image quality. The running complexity results show that the method is quite accurate and can maintain the actual complexity near the target values, with maximum variations around 3 %.

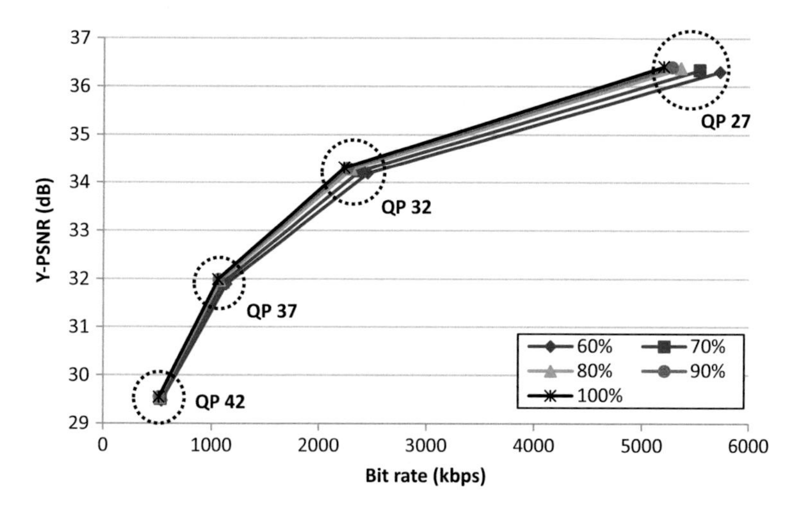

Fig. 5.4 Average R-D performance of the FDCS method

Table 5.1 Average results obtained for the FDCS algorithm

Target complexity (α^{T}) (%)	Running complexity (%)	BD-rate (%)	BD-PSNR (dB)
100	100	–	–
90	89	+1.24	−0.03
80	77	+3.15	−0.09
70	69	+7.59	−0.21
60	63	+12.02	−0.33

The results show that FDCS is capable of adjusting the number of tested coding tree possibilities by increasing or decreasing the maximum coding tree depth used in all treeblocks belonging to a frame. Nevertheless, using a unique, fixed maximum coding tree depth for all treeblocks in the frame harms the R-D efficiency in some cases, especially when the target complexity is set to the lowest value tested (α^{T} = 60 %). In such case, the actual computational complexity reduction achieved is around 37 %, and the BD-rate increase is 12.02 %. Aiming at finding solutions to improve the R-D efficiency of this simple method, the approaches presented in the next sections are proposed.

5.3 Variable Depth Complexity Scaling

The variable depth complexity scaling (VDCS) method [1–3] is based on dynamic constraining of the maximum coding tree depth according to the depth used in previously encoded frames. This method is motivated by the results presented in Fig. 5.2 of Sect. 5.1, which shows the evolution of the maximum coding tree depth used in random co-localised treeblocks of neighbouring frames. Based on those

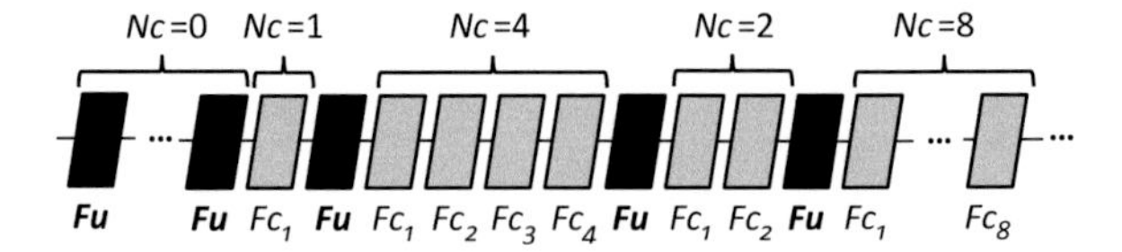

Fig. 5.5 Example of operation of the VDCS strategy

observations and on the strategy presented in Sect. 5.2, a variable depth complexity scaling method was developed.

As in the FDCS method, two types of frames are used in VDCS: unconstrained (Fu) and constrained frames (Fc). Fu frames are also defined as those encoded through the usual process in which all possible coding tree structures are tested. However, different from FDCS, in VDCS the Fc frames are encoded with maximum coding tree depths bound to those used in the most recently encoded Fu frame, taking advantage of the temporal stationarity characteristic of video sequences.

5.3.1 *Algorithm Overview*

As the computational complexity required to encode an Fc frame is smaller than that of an Fu frame, the number of consecutive Fc frames (called here as Nc) is dynamically adjusted as a function of the target computational complexity, as illustrated in the example of Fig. 5.5. Naturally, the larger the value of Nc, the smaller is the computational complexity required to encode the video sequence.

The high-level diagram of the VDCS algorithm is presented in Fig. 5.6. The number of frames which compose the video segment (N) and the complexity reduction ratio ($0\,\% < \alpha^{T} < 100\,\%$) are the two input parameters of the algorithm. Just as in the FDCS method, in the starting phase, CE^{Max} and CE^{T} are computed according to Eqs. (5.1) and (5.2), respectively. While CP^{N} is within the limit imposed by CE^{T}, all frames are encoded as Fu (i.e. Nc=0). Whenever CP^{N} increases beyond CE^{T}, the complexity adjustment phase is activated, and frames start being encoded as Fu or Fc, according to the algorithm's decisions (dashed line box in Fig. 5.6).

The first step in the complexity adjustment phase of the algorithm is the calculation of a new Nc value, which is updated using a proportional control loop. The value of Nc depends on the normalised difference β between CP^{N} and CE^{T}, calculated as in Eq. (5.4). Figure 5.7 shows how Nc is adjusted as a non-linear function of β. The larger the normalised difference between CP^{N} and CE^{T} (horizontal axis), the larger is the decrease or increase of Nc (vertical axis). If the Nc saturation value is reached, the algorithm keeps it constant until a computational complexity increase is allowed.

After the new Nc value is defined, a new Fu frame is encoded, and the maximum coding tree depth used for each treeblock of that frame is stored for future use in a matrix called maximum tree depth map (MTDM). Then, the next Nc-constrained

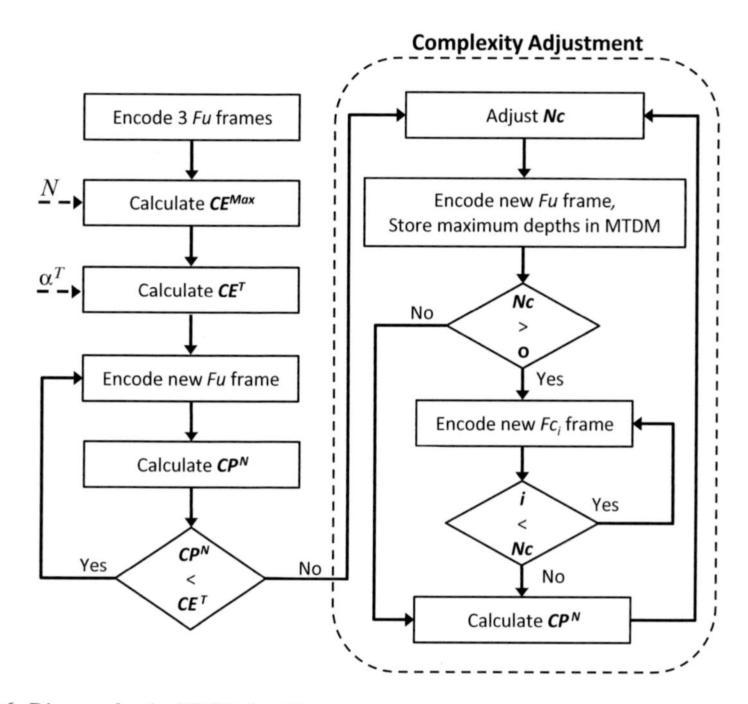

Fig. 5.6 Diagram for the VDCS algorithm

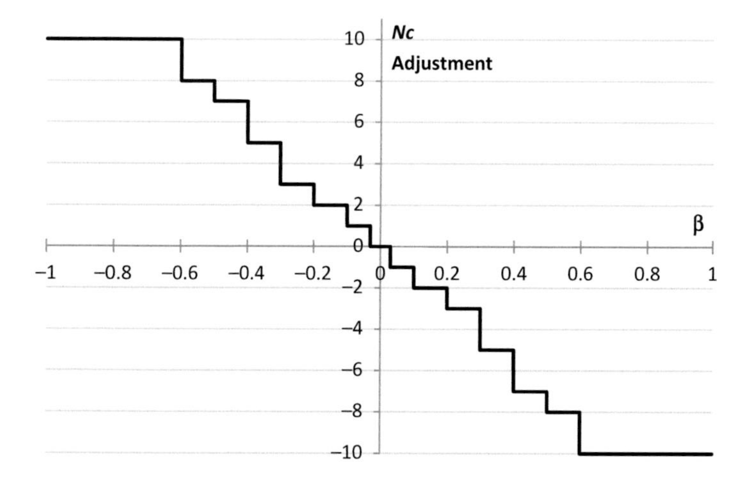

Fig. 5.7 Adjustment of *Nc* value according to β

frames Fc are encoded with the RDO process limited to the values saved in the MTDM, which means that the maximum depths used are upper-bounded by the depths used in the co-localised treeblocks of the most recent Fu frame. CP^N is then once again calculated and Nc is readjusted, if necessary. The CP^N value is computed different from FDCR, according to Eq. (5.5), where CS^{Fi} is the computational complexity spent to encode the *i*th frame of the segment, CS^{Fu} is the complexity used to encode the last Fu frame, CS^{Fcj} is the computational complexity used in the *j*th frame of the last group of constrained frames and NE is the number of frames already encoded in the video segment.

$$\beta = \frac{CE^{T} - CP^{N}}{CE^{T}} \tag{5.4}$$

$$CP^{N} = \sum_{i=1}^{NE} CS^{Fi} + \frac{CS^{Fu} + \sum_{j=1}^{Nc} CS^{Fc_j}}{Nc + 1} \cdot (N - NE) \tag{5.5}$$

5.3.2 *Results for VDCS and Performance Evaluation*

The VDCS method was evaluated using the same experimental setup as that of FDCS, based on HM4, three video sequences comprising 500 frames (*BasketballDrive*, *BQTerrace*, *Cactus*) plus a concatenation of two of them (*BasketballCactus*), the *low delay P* temporal configuration, five target complexities (from 60 to 100 %) and four different QPs (27, 32, 37 and 42). An upper limit (i.e. a saturation value) of ten frames was used for Nc in all experiments. Even though this value can be increased or decreased depending on the system/user requirements, it was found experimentally that larger maximum values for Nc would lead to higher unacceptable R-D efficiency losses.

To illustrate the operation of the algorithm, Fig. 5.8 shows the evolution of the Nc and CP^{N} values along the encoding of the *BasketballDrive* sequence with QP 32. Figure 5.8a shows that large Nc values are used with small target complexities, reaching the saturation value (Nc = 10) when the target is set to 60 %. With a target complexity of 90 %, the maximum Nc value used is 3. It is possible to notice that variations on the CP^{N} value shown in Fig. 5.8b are followed by increases or decreases in the Nc value in Fig. 5.8a. For example, for a target complexity of 60 %, the increase of CP^{N} from frames 20 to 60 caused successive increments on the Nc value for the same frames. Oppositely, in frame 180 the CP^{N} value decreased to a value under the target (dashed line in the chart), which means that more computational resources could be employed in the encoding process. This reflected in a decrease of the Nc value from 10 to 8 in frame 180.

Figure 5.9 shows the relationship between the target complexities defined at the beginning of the encoding process and the actual complexities measured while encoding the video sequences. The dashed line in the graph represents the ideal behaviour of a complexity scaling method. As running complexity results for each tested sequence are around this ideal case, the proposed method is considered quite accurate and capable of scaling computational complexity.

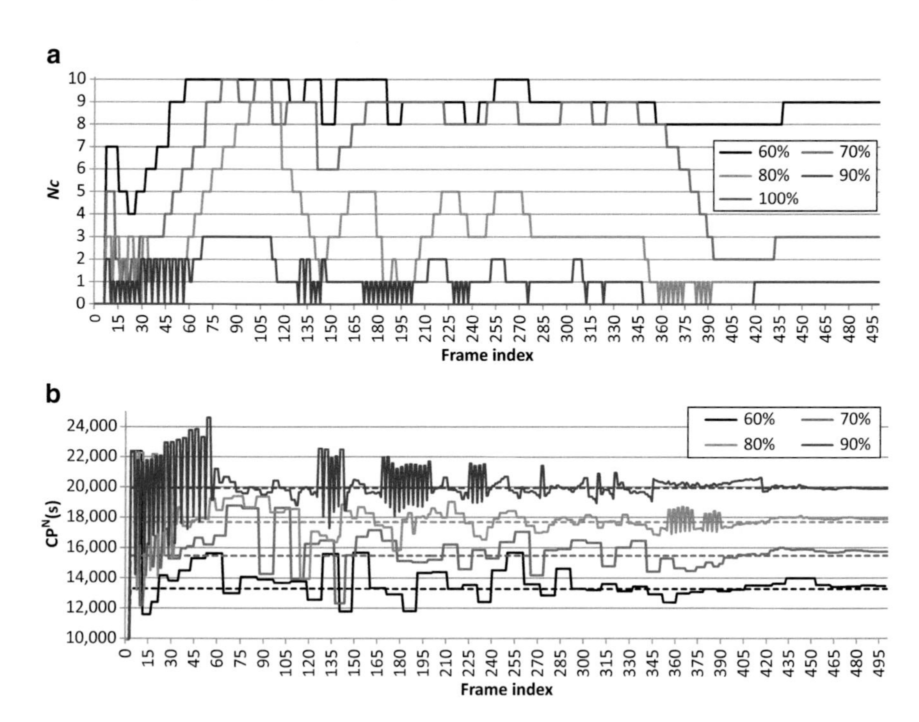

Fig. 5.8 *Nc* and CP^N variations when encoding sequence *BasketballDrive*, QP32

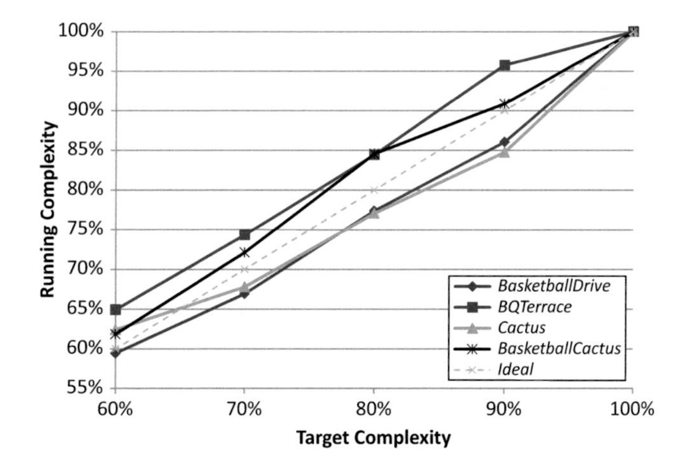

Fig. 5.9 Accuracy of the VDCS complexity scaling method

The encoder R-D efficiency as a function of the computational complexity under different bit rates is presented in Fig. 5.10. As expected, the best results are obtained when no complexity scaling is applied (i.e. $\alpha^T = 100\ \%$). However, when compared to the curves obtained for FDCS (see Fig. 5.4), the results for VDCS are represented by Y-PSNR versus bit rate curves that are much closer to each other, which means

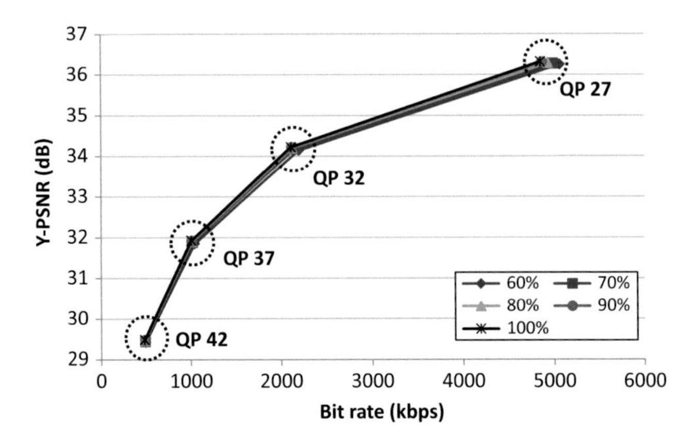

Fig. 5.10 Average R-D performance of the VDCS method

Table 5.2 Average results obtained for the VDCS algorithm

Target complexity (α^T) (%)	Running complexity (%)	BD-rate (%)	BD-PSNR (dB)
100	100	–	–
90	90	+0.58	–0.02
80	80	+1.55	–0.04
70	71	+3.10	–0.09
60	62	+6.29	–0.18

that the complexity scaling algorithm does not affect the compression efficiency significantly, or at least not as much as FDCS, even when small target complexities are set (e.g. $\alpha^T = 60$ %).

Table 5.2 shows average performance results for the VDCS algorithm. The table shows, for each target complexity tested, the average running complexity, the BD-rate and the BD-PSNR values in comparison to the maximum complexity case (100 %). The results show that VDCS is twice more efficient than FDCS, since it achieves similar computational complexity decreases (up to 38 %) at the cost of almost half the BD-rate increase (up to +6.29 %).

Even though to a smaller extent than FDCS, the use of VDCS also incurs in R-D efficiency losses, which also happens mostly in the low target complexities. This is because in such cases the complexity scaling algorithm uses large Nc values, which leads to the use of maximum coding tree depth values that may not match very well the actual image characteristics due to temporal changes accumulated since the last Fu frame encoded. To solve this problem, a new method is presented in the next section, which considers the average motion of each treeblock in the decision of the maximum coding tree depth.

5.4 Motion-Compensated Tree Depth Limitation

Just as FDCS and VDCS, the motion-compensated tree depth limitation (MCTDL) method [4] is based on dynamic constraining the maximum coding tree depth allowed in order to adjust the encoding computational complexity. As previously explained, in VDCS the maximum coding tree depths of Fc frames are constrained according to the MTDM saved while encoding the last Fu frame. However, in video sequences with fast motion segments, this method may decrease the encoding performance, partly due to the fact that in such cases image areas may move away from the position where they were in the last Fu frame before another Fu frame is encoded and the MTDM is updated. As a result, the Fc frames occurring between two consecutive Fu frames are encoded using a MTDM that may not be well matched to the current frame content, with the mismatch becoming worse towards the end of the group of Fc frames. This problem is more severe in cases where the target complexity is small and thus Nc is large.

To solve this problem, MCTDL adds a new step to the complexity adjustment phase of VDCS, updating the MTDM after encoding each Fc frame according to the average motion of each treeblock, effectively motion compensating the MTDM. This compensation of the motion effect is performed using motion information from the previous frame in order to predict the most probable displacement from frame $k-1$ to frame k for the image region corresponding to each CTU.

5.4.1 Algorithm Overview

As in VDCS, the maximum depths saved when encoding an Fu frame are stored in an $n \times m$ matrix (referred from now on as MTDM[n][m]), where n is the number of CTUs in the horizontal dimension and m is the number of CTUs in the vertical dimension of a frame. Simultaneously, a weighted average motion vector (MV) for each CTU is computed and stored in another $n \times m$ matrix (referred from now on as MV[n][m]), where the weights are proportional to the size of each PU inside the CTU, as detailed later in this section. When the first Fc frame which follows an Fu frame is encoded, the values stored in MTDM[n][m] are motion compensated according to the MVs stored in MV[n][m]. Each treeblock in Fc is then encoded with this motion-compensated map of maximum depths. While encoding each Fc frame, new average MVs for each CTU are computed and stored in MV[n][m] in order to update the MTDM to be used in the next Fc frame.

To fully understand the maximum depth map construction procedure, let us explain as an example how the maximum depth to be applied to a certain $CTU^{k-1}_{(o,p)}$ belonging to a frame F_{k-1} is derived. Consider that this CTU average MV is $MV^{k-1}_{(o,p)} = (x, y)$, the reference frame is F_r, the coordinates (o, p) are the CTU line and column in frame F_{k-1} and the coordinates (x, y) are the CTU line and column displacement from frame F_r to frame F_{k-1}. These elements are all shown in Fig. 5.11

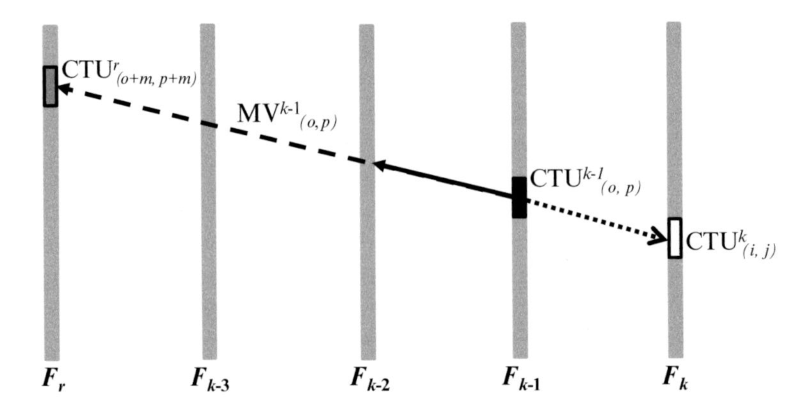

Fig. 5.11 Example of MTDM motion compensation

for better comprehension. Now, let us assume that the motion speed of the group of pixels constituting the CTU is approximately constant. The displacement of the pixels belonging to the example CTU, from frame F_{k-2} to frame F_{k-1}, can be computed by dividing $MV^{k-1}{}_{(o,p)}$ components m and n by the number of frames between frame F_{k-1} and F_r. Still assuming constant motion speed, we can predict where the pixels of $CTU^{k-1}{}_{(o,p)}$ may be located in frame F_k. The solid line arrow in Fig. 5.11 shows the estimated motion displacement from frame F_{k-1} to frame F_{k-2}, and the pointed line arrow shows the predicted motion displacement from frame F_{k-1} to frame F_k. The maximum coding tree depth for each $CTU^k{}_{i,j}$ in frame F_k is then copied from the corresponding position in $MTDM^{k-1}$ and stored into a motion-compensated MTDM for frame F_k, named $MTDM^k$.

Figure 5.12 shows the high-level diagram of the MCTDL algorithm. The operation of this algorithm is very similar to that of VDCS, which was explained in details in Sect. 5.3.1. The starting phase is exactly the same as in VDCS, with the computations of CE^{Max}, CE^T and CP^N performed accordingly to Eqs. (5.1), (5.2) and (5.5), respectively. While CP^N is smaller than CE^T, all frames are encoded as Fu. When this condition ceases to hold, the complexity adjustment phase is activated (portion inside the dashed line in Fig. 5.12), and frames start being encoded as Fu or Fc, according to the algorithm's decisions.

The new operations introduced in MCTDL executed during the complexity adjustment phase are underlined in the diagram of Fig. 5.12. Initially, the calculation of Nc is done as in VDCS, according to the normalised difference between CP^N and CE^T, as in Eq. (5.4). After that, a new Fu frame is encoded, and the maximum coding tree depths are saved in the MTDM. Simultaneously, the encoder records the average weighted MVs belonging to each CTU, so that the MTDM can be motion compensated in the next step.

The average MV is computed as a weighted average of all MVs belonging to the PUs in the CTU. The weights applied to each MV depend on the PU size relative to the CTU size, as shown in Table 5.3. For example, a 64×64 PU has a weight equals to 1, since the number of samples in the PU and in a 64×64 CTU is the same.

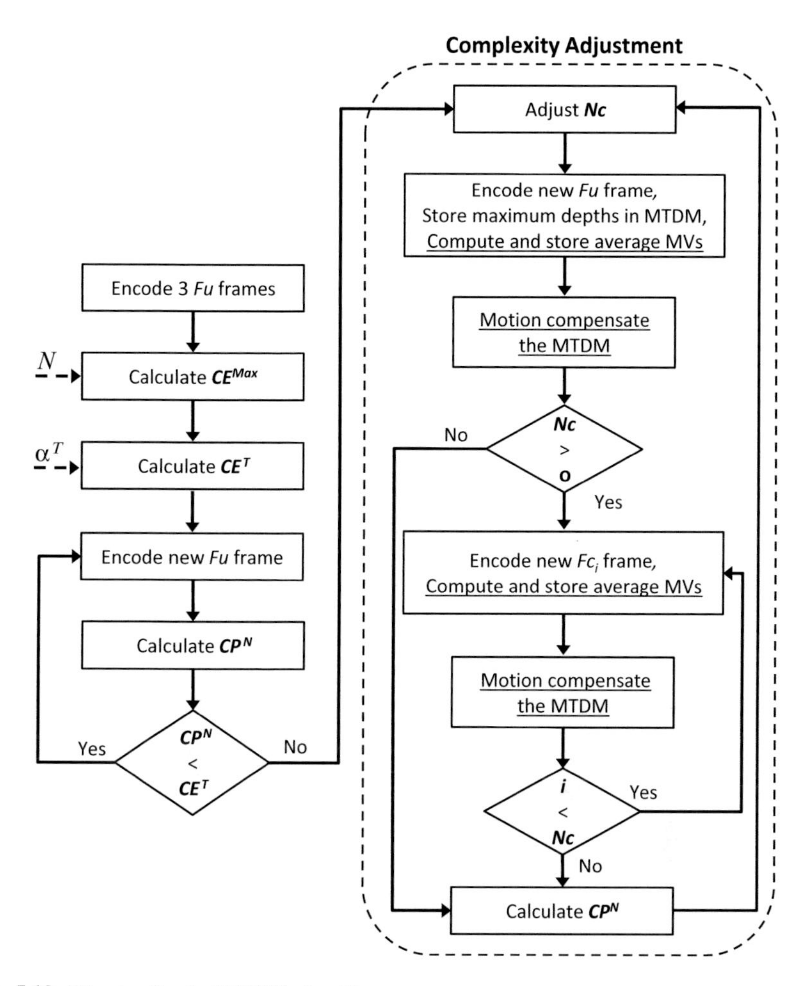

Fig. 5.12 Diagram for the MCTDL algorithm

A 32×32 PU has a weight equals to 1024/4096=0.25, where 1024 is the number of samples in the PU and 4096 is the number of samples in the 64×64 CTU.

If Nc is zero (i.e. no complexity reduction is necessary), the algorithm keeps encoding Fu frames but still computes a new CP^N, and a new Nc after each Fu frame is encoded in order to detect whenever complexity reduction is required. In this case, Nc becomes larger than zero, and Fc frames start being encoded with the maximum depth used in the RDO process limited to the values saved in the MTDM, which are reordered according to the average MVs for each CTU after encoding each Fc frame. By doing that, the encoder keeps an updated estimation of the location where is the CTU corresponding to a determined value in the MTDM.

Different from the VDCS method, in which the MTDM data becomes out of date when large Nc values are used, in this new method the encoder is able to apply a coding tree depth limitation that fits better the characteristics of each image region.

Table 5.3 Weights for average MV calculation

PU size	Number of samples	Weight
64×64	4096	1
64×32	2048	0.5
32×64	2048	0.5
64×16	1024	0.25
16×64	1024	0.25
32×32	1024	0.25
32×16	512	0.125
16×32	512	0.125
32×8	256	0.0625
8×32	256	0.0625
16×16	256	0.0625
16×8	128	0.03125
8×16	128	0.03125
16×4	64	0.015625
4×16	64	0.015625
8×8	64	0.015625
8×4	32	0.0078125
4×8	32	0.0078125

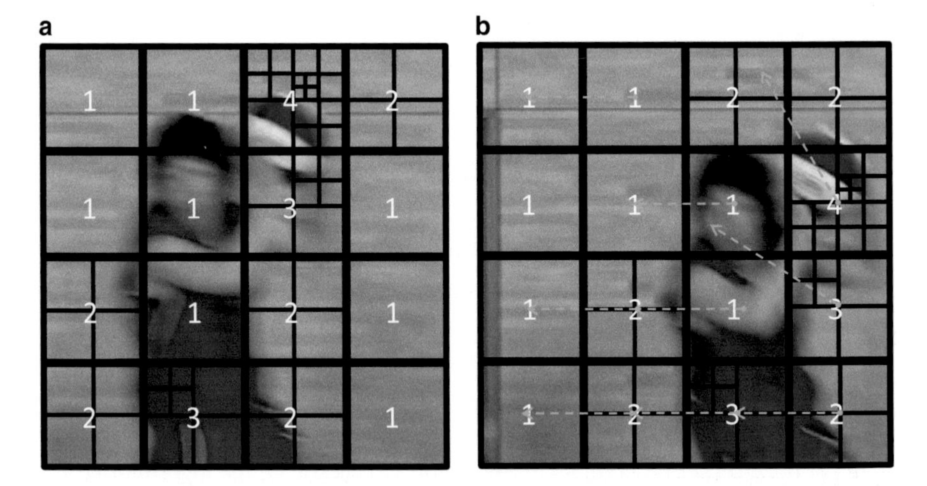

Fig. 5.13 MTDM fragment and average MVs for the (**a**) 58th and (**b**) 59th frames of the *BasketballDrive* sequence (QP 32)

This is illustrated in the example presented in Fig. 5.13, which shows a fragment of the 58th and 59th frames of the *BasketballDrive* video sequence. The maximum coding tree depth is shown for each CTU in the fragments (i.e. the information recorded in the MTDM), and the average MVs calculated as explained before are shown in Fig. 5.13b for each inter-predicted CTU. It is possible to notice in the figure that motion compensating the MTDM according to the average MV yields a good estimation of the maximum coding tree depth for the next frame.

5.4.2 Results for MCTDL and Performance Evaluation

The same setup used in the evaluation of FDCS and VDCS was used for performance assessment of the MCTDL method.

Figure 5.14 shows a graph with the actual complexities obtained after encoding the video sequences as a function of the target complexities. The figure presents results for QP 32, but other values were also tested and show similar behaviour. The dashed line represents the target complexity, and the solid lines represent each video tested. As each video sequence presents results that are close to the ideal, it is possible to conclude that the proposed method is accurate and capable of scaling computational complexity.

Table 5.4 presents average results for MCTDL in terms of running complexity, BD-rate and BD-PSNR differences for all target complexities tested in comparison to the case in which no complexity scaling is applied, i.e. when target complexity is 100 %. An increase in the bit rate was observed in all cases, but especially when the target

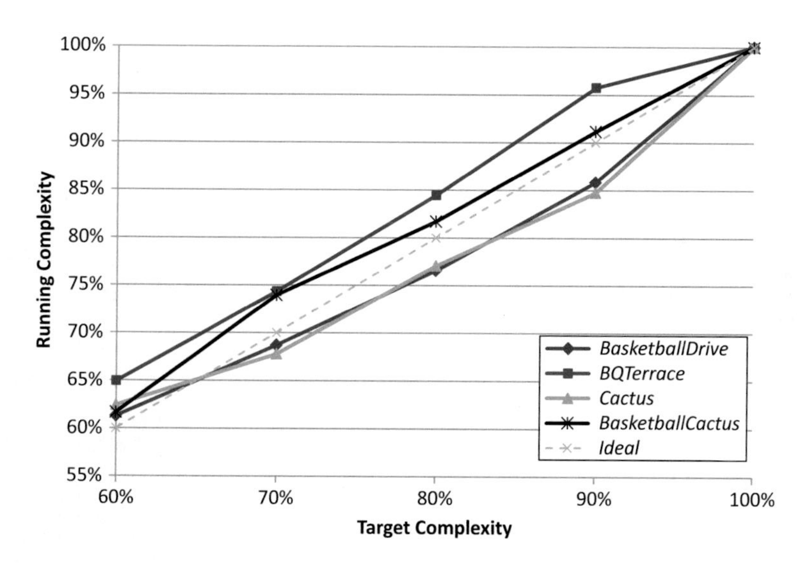

Fig. 5.14 Accuracy of the MCTDL complexity scaling method

Table 5.4 Average results obtained for the MCTDL algorithm

Target complexity (α^T) (%)	Running complexity (%)	BD-rate (%)	BD-PSNR (dB)
100	100	–	–
90	90	+0.54	–0.02
80	80	+1.46	–0.05
70	71	+3.22	–0.10
60	61	+5.42	–0.16

complexity (α^{T}) is set to 60 %. However, in comparison to VDCS and FDCS, the MCTDL method yields better R-D results for all target complexities. In the lowest target complexity case (α^{T}=60 %), an average complexity reduction of 39 % is achieved at the cost of a BD-rate increase of 5.42 %.

5.5 Coding Tree Depth Estimation

Despite providing a fair complexity scaling without significantly decreasing compression efficiency, the MCTDL method (as well as VDCS) uses only the temporal correlation between neighbouring frames in order to determine the maximum coding tree depth tested for each CTU. If spatial correlation was also used combined with temporal data, the R-D performance achieved when complexity scaling is enabled could be potentially improved.

This section presents a coding tree depth estimation (CTDE) method [5, 6], which is based on the method presented in Sect. 5.4 and allows estimating the best maximum coding tree depth for a CTU based on both spatial and temporal correlations observed in neighbouring CTUs located in the same and previous frames. As explained before, the computational complexity scaling approaches used in the previously presented methods rely on the fact that the maximum coding tree depth tends to be constant in co-localised areas of adjacent frames, as experimentally verified. For the CTDE method, a set of experiments were conducted in order to analyse how frequently a certain CTU is encoded with the same or smaller maximum coding tree depth than its spatially neighbouring CTUs (top, left and top-left CTUs). The results of such experiments are presented in Table 5.5 and show that in most cases the CTUs surrounding a given CTU are encoded with maximum depths that are equal to or exceed the maximum depth of that CTU. As all coding tree depths from 0 to n are tested through RDO when a depth n is selected as maximum, no R-D efficiency losses would be observed if a depth greater than the one that should be used for encoding that CTU was selected as maximum.

Based on these observations and considering the complexity scaling scheme of VDCS and MCTDL, in CTDE the maximum coding tree depth allowed for each CTU is decided, taking into consideration the type of frame (Fc or Fu), as well as the maximum depths used in the temporally and spatially neighbouring CTUs.

Table 5.5 Maximum depths used in neighbouring CTUs

Spatially neighbouring CTU	Greater or equal depth (%)	Smaller depth (%)
Top	83.47	16.53
Left	82.24	17.76
Top-left	91.84	8.16

5.5.1 Algorithm Overview

Let $CTU^k_{i,j}$ be a coding tree block located at position i, j of a frame with index k. If k is an Fu frame, the CTU is encoded with no complexity limitation, which means that the maximum coding tree depth possible is allowed. If k is an Fc frame, the maximum coding tree depth allowed is defined to be the largest of the maximum coding tree depths used at the:

1. Left-side neighbouring CTU, i.e. $CTU^k_{(i,j-1)}$
2. Top neighbouring CTU, i.e. $CTU^k_{(i-1,j)}$
3. Top-left neighbouring CTU, i.e. $CTU^k_{(i-1,j-1)}$
4. Co-localised CTU from the previous frame, i.e. $CTU^{k-1}_{(i,j)}$
5. Motion-compensated CTU from the previous frame, i.e. $CTU^{k-1}_{(o,p)}$

Except for the last value listed above, which requires some processing to be obtained, all values are straightforward to obtain by simply storing coding tree depths used in each CTU. Two MTDMs are used to store values corresponding to CTUs in the current and previous frames (k and $k-1$ in display order): $MTDM^k$ and $MTDM^{k-1}$, respectively. A third MTDM is used to store values of the motion-compensated CTUs from the previous frame: $CMTDM^k$.

The CTDE method follows a complexity scaling scheme very similar to MCTDL, so that the high-level diagram previously presented in Fig. 5.12 is slightly modified for CTDE, as shown in Fig. 5.15. In Fig. 5.15, the steps which are incorporated into CTDE or modified from MCTDL are underlined. While a frame is encoded (both Fu and Fc) in the complexity adjustment phase, the maximum coding tree depths are stored in $MTDM^k$, and the average MVs are computed and saved. Then, after encoding the whole frame, $CMTDM^k$ is created by motion compensating the values in $MTDM^k$ (as explained in Sect. 5.4), which is finally copied to $MTDM^{k-1}$.

The main difference between the CTDE and MCTDL consists in the way the maximum coding tree depth allowed for each CTU is computed. As this process is not depicted in Fig. 5.15, since it is a part of the instruction to encode a new frame ("Encode new Fu frame" and "Encode new Fc_i frame" lines in Fig. 5.15), it is detailed in the pseudocode presented in Fig. 5.16, which is executed for each frame (both Fu and Fc).

In an Fu frame, all coding tree depths are allowed, so that the encoder sets the variable *max_depth_allowed* to the maximum coding tree depth possible (lines 04–05, in Fig. 5.16), which varies from 1 to 4, according to the encoder configuration used. In an Fc frame, the *max_depth_allowed* value is decided for each CTU by taking the maximum value among $MTDM^k_{(i,j-1)}$, $MTDM^k_{(i-1,j)}$, $MTDM^k_{(i-1,j-1)}$, $MTDM^{k-1}_{(i,j)}$ and $CMTDM^k_{(i,j)}$ (lines 06–08), which correspond to the five depths listed in the first paragraphs of this section.

To allow further computational complexity reduction, in cases with very small target complexities, the CTDE method allows decreasing the maximum coding tree depth by one additional unit if the maximum value for Nc is already achieved but complexity is still above the target. This is shown in lines 09–10 in Fig. 5.16.

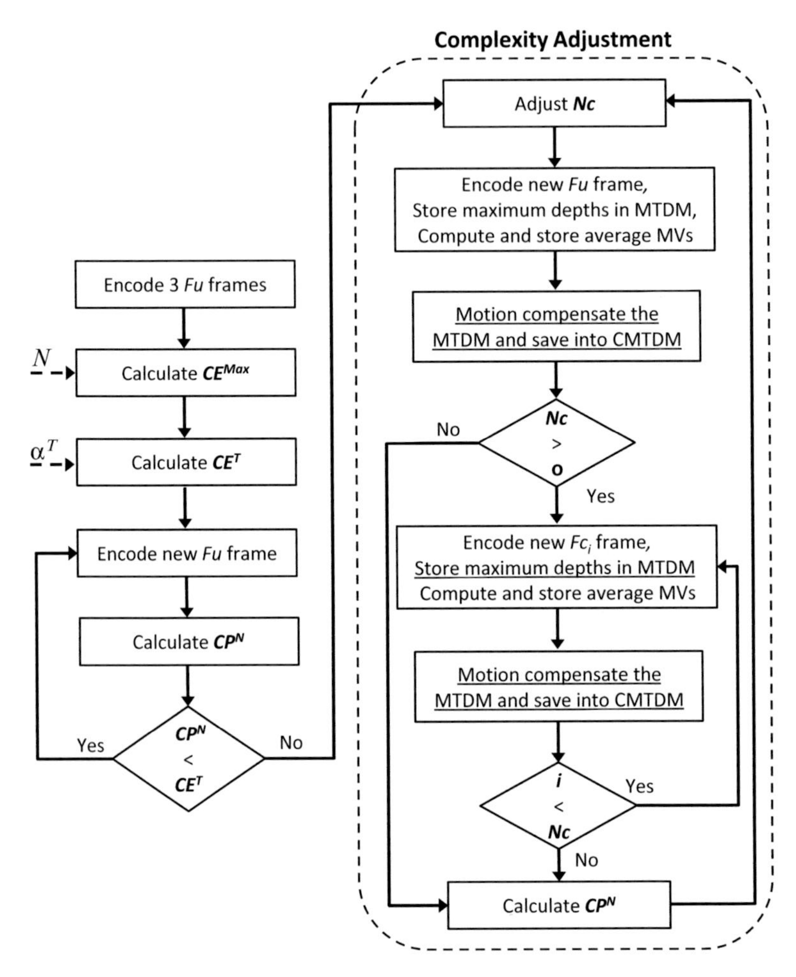

Fig. 5.15 Diagram for the CTDE algorithm

If Nc reaches its maximum value (*MAX_Nc*), the computational complexity can be further reduced by decreasing *max_depth_allowed* by one unit. Even though *MAX_Nc* can be theoretically set to any value, increasing it too much leads to large long-term R-D efficiency losses. To avoid this problem, and following the results of empirical tests, the value of *MAX_Nc* used in the experiments presented in the next section was set to half the video frame rate.

The CTU is finally encoded according to *max_depth_allowed* (line 11), and the maximum coding tree depth used to encode it after testing all tree possibilities is stored in MTDM^k (line 12) for use in the decision of maximum depths for the next CTUs.

```
01   for each CTU in a frame k
02     i ← CTU line position
03     j ← CTU column position
04     if k is an Fu frame
05       max_depth_allowed ← maximum depth possible
06     else
07       max_depth_allowed ← max(MTDM^k_(i-1,j), MTDM^k_(i,j-1),
08                               MTDM^k_(i-1,j-1), MTDM^k-1_(i,j), CMTDM^k_(i,j))
09       if Nc = MAX_Nc
10         max_depth_allowed ← max_depth_allowed - 1
11     encode CTB limiting depth to max_depth_allowed
12     MTDM^k_i,j ← maximum depth used to encode current CTU
13   end for
```

Fig. 5.16 Pseudocode for the maximum coding tree depth decision in CTDE

5.5.2 Results of CTDE and Performance Evaluation

The CTDE method was evaluated by having its complexity scaling accuracy and R-D efficiency measured under specific target complexities. The algorithm was implemented in the HM encoder—version 8.2 (HM8.2)—and evaluated with six video sequences (*BQMall*, *BQTerrace*, *Cactus*, *vidyo1*, *BasketballDrive*, *Traffic*), all of which are detailed in Appendix A. The encoder performance was evaluated under five target complexities: 60, 70, 80, 90 and 100 %. The *low delay P* temporal configuration [9] was used in all tests.

Figure 5.17 shows a chart with the average complexities obtained after encoding the video sequences as a function of the target complexities. The dashed line in the graph represents the target complexity, while the other lines represent the actual running complexity for various test video sequences. As it can be seen, the actual running complexity for each tested sequence is close to the target, thus showing that CTDE is accurate and capable of scaling computational complexity to within a tight interval around the desired value. In the worst case ($\alpha^{T}=60$ %), the difference between target and actual running computational complexity was around 3 %.

Concerning the video encoding performance, Table 5.6 presents average results when the CTDE method is used. Besides the coding performance indicator variations (i.e. BD-rate increases and BD-PSNR decreases), the table also shows average results for running complexity, considering all videos and QPs are tested. An increase in the bit rate was observed in all sequences coded at low complexity points, especially when the target complexity is set to 60 %. However, in comparison to all methods presented so far in this book, CTDE produced the best R-D

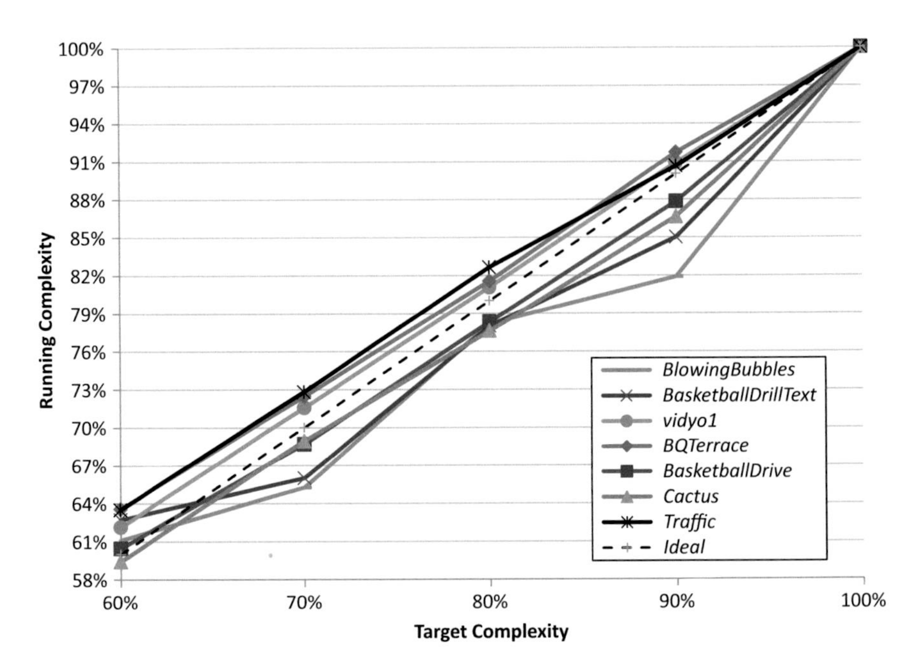

Fig. 5.17 Accuracy of the CTDE complexity scaling method

Table 5.6 Average results obtained for the CTDE algorithm

Target complexity (α^{T})	Running complexity	BD-rate (%)	BD-PSNR (dB)
100 %	100 %	–	–
90 %	89 %	+1.05	−0.03
80 %	80 %	+1.62	−0.04
70 %	72 %	+3.84	−0.10
60 %	63 %	+4.74	−0.13

efficiency results for all target complexities, with encoding efficiency closer to the values obtained when no complexity scaling is applied. In the worst case ($\alpha^{\mathrm{T}} = 60$ %), the BD-rate increase was 4.74 %, and the decrease in BD-PSNR was 0.13 dB.

5.6 Constrained Coding Units and Prediction Units

As the experiments presented in Chap. 4 have shown, CUs are the frame partitioning structures which allow the largest reductions in computational complexity when constrained partition definition procedures are used. However, the same experiments have also exposed that constraining CUs in order to decrease computational complexity incurs in the largest costs in terms of R-D efficiency loss in comparison to PUs and TUs (see Tables 4.6 and 4.7).

According to the experiments presented in Chap. 4, the frame partitioning structure that yields the second largest reductions in computational complexity when constrained are PUs (see Table 4.6). They are also the structures which presented the second smallest ratios between BD-rate and computational complexity decrease (see Table 4.7), which means that constraining PUs is more advisable than constraining CUs if maintaining the R-D efficiency is a crucial preoccupation in the system implementation.

Based on that analysis and on the methods for computational complexity scaling previously presented, a method based on constrained coding units and prediction units (CCUPU) [7, 8] is proposed in this section. The goal of this method is to allow adjusting the HEVC encoding structures in order to achieve a dynamic scaling of the computational complexity beyond levels achieved by the previous approaches.

5.6.1 Rate-Distortion-Complexity Relationship in CUs and PUs

A more extensive analysis of the R-D-C efficiency of different CU and PU configurations was performed in order to assist the development of the CCUPU method. This analysis was necessary mainly because the constraining of PUs performed in the experiments of Chap. 4 was based on simply enabling or disabling the use of AMP, which is the only parameter regarding the configuration of PUs available in the HM configuration. For the experiments described in this section, a configuration in which PU sizes smaller than $2N\times 2N$ are disabled was created through modifications of the HM code. This $2N\times 2N$-only configuration was used in the experiments and compared with the unmodified HM encoder. A subset of ten video sequences (*BlowingBubbles*, *RaceHorses1*, *BasketballDrillText*, *PartyScene*, *SlideShow*, *vidyo1*, *ParkScene*, *BasketballDrive*, *NebutaFestival* and *Traffic*) was selected for tests from the CTC document. Their characteristics are detailed in Appendix A.

The ten video sequences were encoded using four different QPs (27, 32, 37 and 42) and the *low delay* P temporal encoder configuration while varying two parameters: the maximum coding tree depth allowed for the CTUs and the possibility of splitting a CU into PUs smaller than $2N\times 2N$ for inter-frame prediction. Table 5.7 presents the eight configurations tested. Table 5.8 shows comparative results in terms of BD-rate and computational complexity for the different configurations presented in Table 5.7. In the first three rows of Table 5.8, those configurations in which only the maximum coding tree depth varies (*Cfg. 1* to *Cfg. 4*) are compared, while the remaining four rows compare results when only the possibility of using PUs smaller than $2N\times 2N$ varies.

The results in Table 5.8 show that when the maximum coding tree depth allowed is reduced by one unit, an average increase of 3.33 % in the BD-rate and a computational complexity decrease of 41.3 % are observed. On the other hand, turning off the possibility of using PUs smaller than $2N\times 2N$ in inter-frame prediction results in an average BD-rate increase of 0.93 % and a computational complexity decrease of 43.5 %, which seems much more profitable from the point of view of R-D-C trade-offs.

Table 5.7 CU and PU configurations tested

Configuration	Max. coding tree depth	PU smaller than $2N \times 2N$
Cfg. 1	4	Yes
Cfg. 2	3	
Cfg. 3	2	
Cfg. 4	1	
Cfg. 5	4	No
Cfg. 6	3	
Cfg. 7	2	
Cfg. 8	1	

Table 5.8 Comparison between encoding configurations

Comparison	BD-BR (%)	ΔComp (%)	BD-BR/ ΔComp ($R_{i,j}$)	Avg. BD-BR (%)	Avg. ΔComp (%)
Cfg. 2 vs. *Cfg. 1*	+1.10	−28.4	0.0373	+3.33	−41.3
Cfg. 3 vs. *Cfg. 2*	+2.05	−45.5	0.0451		
Cfg. 4 vs. *Cfg. 3*	+6.84	−49.9	0.1371		
Cfg. 5 vs. *Cfg. 1*	+1.13	−44.1		+0.93	−43.5
Cfg. 6 vs. *Cfg. 2*	+1.09	−49.3			
Cfg. 7 vs. *Cfg. 3*	+0.68	−43.4			
Cfg. 8 vs. *Cfg. 4*	+0.81	−37.1			

R-D costs and computational complexity were jointly analysed by computing the ratio $R_{i,j}$ between the BD-rate increase and the decrease in computational complexity ΔComp when the maximum coding tree depth is reduced from i to j (without changing the possibility of using PUs smaller than $2N \times 2N$). By calculating the ratio, it is possible to compare the configurations in order to detect which one incurs in the largest impact on compression efficiency per computational complexity saving. The results are presented in the fourth column of Table 5.8 and show that the ratio $R_{i,j}$ is smaller when decreasing coding tree depths of higher values (from *Cfg. 1* to *Cfg. 2* and from *Cfg. 2* to *Cfg. 3*), even though ΔComp is moderate in such cases. On the other hand, when decreasing the coding tree depth of lower values (from *Cfg. 3* to *Cfg. 4*), the ratio $R_{i,j}$ increases significantly (0.1371), but ΔComp also increases. These results show that, after constraining the PU size, the best option for decreasing computational complexity is to restrict the use of large coding tree depths.

5.6.2 Encoding Constrained CTUs

TheCCUPU method scales the computational complexity by using a two-level constraining scheme. As suggested by the experiments presented in the previous section, a better R-D efficiency is achieved by the encoder if limiting PU shapes is performed prior to limiting coding tree depths. Thus, in the first constraining level, the number of CUs that are split into PUs smaller than $2N \times 2N$ for inter-prediction is adjusted to scale the computational complexity. Then, if the first constraining level is not enough to achieve the target complexity, the number of CTUs that can be encoded at any coding tree depth is also adjusted. The amount of CTUs constrained in the first and second levels is controlled by two parameters: N_c^1 and N_c^2, respectively.

CTUs constrained with the first parameter (N_c^1) are called PU-constrained CTUs, while CTUs constrained with the second parameter (N_c^2) are called CU-constrained CTUs. Notice, however, that CU-constrained CTUs are also PU-constrained CTUs, since both parameters are used for complexity scalability in these cases.

When the first parameter is used for complexity constraining, the encoder disables PU shapes smaller than $2N \times 2N$ in those CTUs that most probably yield low R-D costs according to information of their co-localised CTUs in the previous frame, as explained in the next section. Once the number of PU-constrained CTUs reaches the number of CTUs in a frame, the second constraining parameter is changed to further reduce computational complexity. In this case, the method adjusts the number of CTUs in which the maximum coding tree depth is decided by taking into account previous encoding decisions, based on the fact that spatial and temporal neighbouring CTUs tend to present the same or similar maximum coding tree depths, as discussed in the previous methods presented in this chapter.

For the CCUPU method, these characteristics were further exploited by analysing additional spatial neighbouring CTUs of a certain CTU. Besides the top, left and top-left CTUs, which were already analysed in Table 5.5, statistics for the top-right CTU and the co-localised CTU in the previous frame are shown in Table 5.9. The results show that in most cases the CTUs surrounding the CTU to be encoded and its co-localised CTU in the previous frame are encoded with maximum depths that are equal to or exceed the maximum depth of the CTU under consideration.

As in the case of unconstrained CTUs, the proposed method also uses RDO to decide the best coding tree configuration in CU-constrained CTUs. However, instead of using a fixed maximum depth for every CTU, this value varies from one CTU to another according to spatio-temporal correlation. In CTUs with smaller maximum depths, the number of coding tree possibilities is smaller, and thus the computational cost of finding the best tree configuration is reduced.

Table 5.9 Maximum depths used in neighbouring CTUs

Spatially neighbouring CTU	Greater or equal depth (%)	Smaller depth (%)
Co-localised	93.14	6.86
Top	83.47	16.53
Left	82.24	17.76
Diagonal top-left	91.84	8.16
Diagonal top-right	86.33	13.67

Based on this analysis, the maximum coding tree depth *max_depth* used to encode a CU-constrained $CTU^{k}_{(i,j)}$ located at position (i, j) of frame k is defined as the largest of the maximum coding tree depths used at the:

1. Left-side neighbouring CTU, i.e. $\mathrm{CTU}^{k}_{(i,j-1)}$
2. Top neighbouring CTU, i.e. $\mathrm{CTU}^{k}_{(i-1,j)}$
3. Top-left neighbouring CTU, i.e. $\mathrm{CTU}^{k}_{(i-1,j-1)}$
4. Top-right neighbouring CTU, i.e. $\mathrm{CTU}^{k}_{(i-1,j+1)}$
5. Co-localised CTU from the previous frame, i.e. $\mathrm{CTU}^{k-1}_{(i,j)}$

The five values above are obtained directly from stored information of CTUs already encoded in the current and previous frame and saved in the MTDM structures used in VDCS, MCTDL and CTDE methods. Since all tree depths, from 1 to *max_depth*, are evaluated when the RDO process is active, no loss of R-D efficiency is observed if a depth greater than the one that should be used for encoding the current CTU is selected as maximum.

5.6.3 Algorithm Overview

The pseudocode for the CCUPU method is presented in Fig. 5.18. The method starts by calculating the target time $T_{\mathrm{t}}^{\mathrm{GOP}}$ for a GOP, which is needed throughout the encoding process to adjust the number of constrained CTUs at each level of the constraining scheme. This adjustment is performed according to the difference between the time used to encode the previous GOP and the target encoding time for a GOP. For real-time encoding and typical GOP sizes (e.g. 16 frames or smaller) and frame rates (e.g. 50 frames per second), adjustments at GOP granularity are at sub-second scale.

The target time $T_{\mathrm{t}}^{\mathrm{GOP}}$ for a GOP is calculated following Eq. (5.6), where *nFR* is the number of frames in a GOP and T_{t} is the target time to encode a frame, both of which are inputs to the complexity scaling algorithm. The parameter T_{t} is specified by the user or given by the encoding system according to the current load level of the processing elements (e.g. percentage of CPU in use) or the device battery status, and it represents the average maximum time per frame that the encoder is expected to use. As the GOP encoding time depends on the number of frames composing it, T_{t} is used as an input for further calculation of $T_{\mathrm{t}}^{\mathrm{GOP}}$. The calculation of $T_{\mathrm{t}}^{\mathrm{GOP}}$ is shown in line 02 in the pseudocode for the complexity scalability method presented in Fig. 5.18.

$$T_{\mathrm{t}}^{\mathrm{GOP}} = T_{\mathrm{t}} \cdot n\mathrm{FR} \tag{5.6}$$

At least one GOP in a sequence must be encoded without any restriction in order to allow the encoder to detect if complexity adjustment is needed or not. The first GOP is encoded with only unconstrained CTUs (i.e. the full RDO process is applied to all CTUs in its frames), as shown in lines 03–08 in Fig. 5.18, and the time spent

```
01  start
02  calculate T_t^GOP
03  start a new GOP
04  for each j from 0 to nFR
05    for each i from 0 to nCTU
06      mark CTU i as unconstrained
07      encode CTU i
08    if last frame go to line 01
09  calculate T_e^GOP
10  calculate α^GOP
11  if N_c^1 < nCTU
12    calculate new N_c^1
13  else
14    calculate new N_c^2
15  start a new GOP
16  for each j from 0 to nFR
17    sort CTUs in ascending order of R-D cost
18    for each i from 0 to nCTU
19      if i < N_c^k
20        mark CTU i as constrained
21      else
22        mark CTU i as unconstrained
23      encode CTU i
24    if last frame go to line 01
25  go to line 09
```

Fig. 5.18 Pseudocode for the CCUPU method

to encode each frame is saved. Based on the encoding time of the first GOP, the method starts adjusting (if necessary) the computational complexity in order to achieve the target processing time. If the time spent to encode the frames of the previous GOP is larger than the target, then the number of complexity-constrained CTUs in each frame is increased in the next GOP. Otherwise, it is decreased. This process is repeated for every GOP.

The ratio α^{GOP} between the encoding time of the previous GOP and the target time is computed as shown in Eq. (5.7), where $T_{\mathrm{e}}^{\mathrm{GOP}}$ is the actual time elapsed when encoding the previous GOP. The value of α^{GOP} is used as a parameter in the proportional control loop shown in Eq. (5.8), which adjusts the number N_{c}^{k} of constrained CTUs per frame in the next GOP. In N_{c}^{k}, k represents which of the two complexity-constraining levels is used. The method starts scaling the computational complexity by adjusting the number of CTUs which allow PUs smaller than $2N\times2N$. This number is represented as N_c^1 in the pseudocode of Fig. 5.18. Once N_c^1 reaches its limit (i.e. the number of CTUs in a frame), the second parameter in the scheme is used, which is the number of CTUs with constrained maximum coding tree depth, represented as N_c^2. This step is shown in lines 09–14 of Fig. 5.18.

Constrained CTUs are distributed according to the R-D costs of co-localised CTUs in the previous frame as follows. After encoding each CTU, the R-D costs of their CUs are summed up to obtain the CTU R-D cost. When every CTU is finally encoded, the CTU R-D costs are sorted in ascending order (line 17 of Fig. 5.18). CTUs at the beginning of the sorted list are those with lowest R-D cost and are thus less probable to yield high bit rates or distortions if constrained encoding is applied to them in the next frame. On the contrary, CTUs at the end of the list are those with highest bit rates and/or distortions, requiring an unconstrained encoding process to improve R-D performance. Therefore, in the proposed method, the N_c^k CTBs with smallest R-D cost in the sorted list are encoded as constrained, while the remaining CTUs are encoded as unconstrained. This process is described in lines 18–23 of Fig. 5.18.

$$\alpha^{\mathrm{GOP}} = \frac{T_{\mathrm{e}}^{\mathrm{GOP}}}{T_{\mathrm{t}}^{\mathrm{GOP}}} \tag{5.7}$$

$$N_{\mathrm{c}}^{k} \leftarrow \alpha^{\mathrm{GOP}} \cdot N_{\mathrm{c}}^{k} \tag{5.8}$$

5.6.4 Results for CCUPU and Performance Evaluation

The performance of the CCUPU method was evaluated by measuring the trade-off between processing time and R-D efficiency. The same video sequences listed in Sect. 5.6.1 were used in the tests, as well as the *low delay P* temporal encoder configuration, HM—version 9.2 (HM9.2) and QPs 27, 32, 37 and 42. For simulation purposes, the target processing times are defined as a percentage of the fully unconstrained case, i.e. when complexity scalability is not used. The tested target times were calculated corresponding to 10, 20, 30, 40, 50, 60, 70, 80 and 90 % of the average processing time per frame when no complexity constraint is used for a sequence.

Complexity scalability accuracy was measured by comparing the target processing times to the actual times spent for encoding the video sequences. Table 5.10 shows average results for the sequences with different spatial resolutions, encoded under the different target complexities. In Table 5.10, TT corresponds to the target time per frame, and ET is the average elapsed time per frame, both of which are presented as percentages of the time when no complexity scaling is applied. Both BD-BR (which stands for the BD-rate) and BD-PSNR values were calculated using the original HM encoder with no complexity constraining as reference. Average results considering the five resolutions tested are presented in Table 5.11.

Table 5.10 Average complexity and R-D results for CCUPU considering five different spatial resolutions

	416×240			832×480			1280×720			1920×1080			2560×1600		
TT (%)	ET (%)	BD-BR (%)	BD-PSNR (dB)	ET (%)	BD-BR (%)	BD-PSNR (dB)	ET (%)	BD-BR (%)	BD-PSNR (dB)	ET (%)	BD-BR (%)	BD-PSNR (dB)	ET (%)	BD-BR (%)	BD-PSNR (dB)
90	96	0.04	0.00	95	0.05	0.00	92	0.01	0.00	91	0.02	0.00	91	0.04	0.00
80	86	0.40	−0.02	82	0.21	−0.01	82	0.06	0.00	81	0.10	0.00	82	0.41	−0.02
70	73	0.95	−0.04	71	0.47	−0.02	71	0.34	−0.03	71	0.24	−0.01	71	0.27	−0.01
60	62	1.49	−0.06	61	0.79	−0.04	59	0.81	−0.07	61	0.45	−0.01	61	0.44	−0.02
50	53	2.12	−0.09	51	1.57	−0.07	50	1.28	−0.10	51	0.79	−0.03	51	0.64	−0.02
40	45	6.83	−0.27	41	7.48	−0.33	42	2.00	−0.15	41	1.79	−0.06	42	1.82	−0.06
30	35	19.9	−0.75	31	19.2	−0.81	32	7.47	−0.53	31	6.00	−0.19	32	5.22	−0.18
20	27	33.5	−1.23	22	32.6	−1.30	23	22.5	−1.41	21	14.5	−0.43	22	10.1	−0.35
10	24	40.7	−1.47	15	51.8	−1.92	14	55.9	−2.94	13	31.4	−0.86	14	17.6	−0.61

Table 5.11 Average complexity and R-D results for CCUPU

TT (%)	ET (%)	BD-BR (%)	BD-PSNR (dB)
90	93	0.03	0.00
80	83	0.23	–0.01
70	72	0.46	–0.02
60	61	0.80	–0.04
50	51	1.28	–0.06
40	42	3.98	–0.18
30	32	11.55	–0.49
20	23	22.64	–0.94
10	16	39.47	–1.56

A comparison between the values at the TT and ET columns in Tables 5.10 and 5.11 leads to the conclusion that the algorithm is capable of scaling computational complexity quite accurately, since the target and elapsed time are very close in most cases. This is clearer in Fig. 5.19, which presents charts with TT values on the horizontal axis and ET results on the vertical axis for all the video sequences with four different QPs and nine target times. These figures show that for almost all the 360 tests executed in these experiments, the proposed method was able to scale processing times with high accuracy, i.e. with small differences between TT and ET. The only TT value which was not achieved for all video sequences was the 10 % case, setting the lower limit of the achievable computational complexity to values between 10 and 20 % of the unconstrained case. For this reason, the discussion presented in the next paragraphs will focus on the results obtained when TT is set between 20 and 90 %.

According to the results presented in Table 5.10 and Fig. 5.19, the only cases which do not accurately scale computational complexity are the 416×240 video sequences. More specifically, the method was not able to scale computational complexity for the *RaceHorses1* sequence under 43 % for QP 27, 28 % for QP 32 and 22 % for QP 37, as shown in Fig. 5.19a–c. This happens most likely because this sequence is the one with smallest spatial and temporal resolution among all the video sequences tested, resulting on an encoding process which is much faster in comparison to the other video sequences. As the encoding time is already small for the *RaceHorses1* sequence, the proposed method is not able to scale it down to values around 15 %, as in the other sequences. As Table 5.10 shows, the average ET results for 416×240 sequences do not scale under 24 %, and only when TT is set to values above 40 % the method yields appropriate ET results. We can conclude from this analysis that the proposed method is more accurate for scaling the complexity of high-resolution video, which is in fact the target of the HEVC standard and the case where it is more important to reduce and scale encoding complexity due to its larger absolute value.

Concerning the effect of the complexity scaling on encoding efficiency, Table 5.11 shows that the R-D performance tends to decrease significantly only when small target times are set. This happens because small target times incur in the

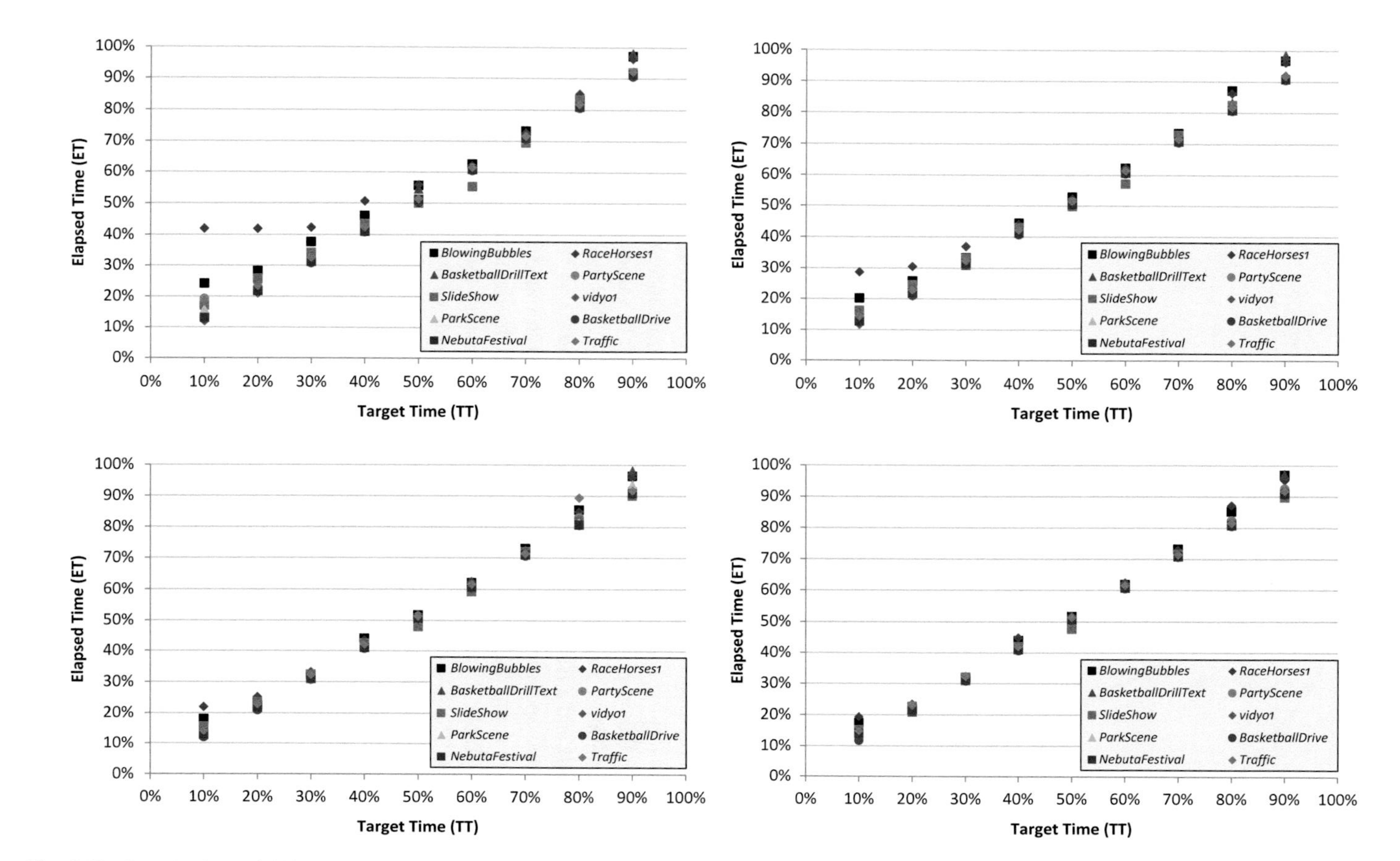

Fig. 5.19 Complexity scalability accuracy for ten video sequences using QP 27 (**a**), QP 32 (**b**), QP 37 (**c**) and QP 42 (**d**)

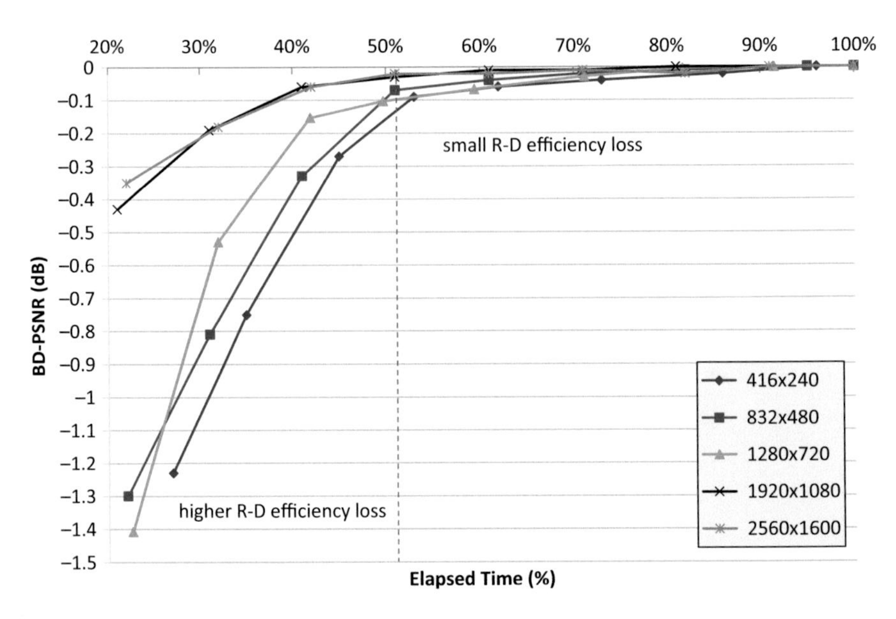

Fig. 5.20 Effect of computational complexity scalability over the R-D performance

use of a larger number of constrained CTUs, which are not optimally encoded through full RDO. On the one hand, for most target times these variations are negligible, more specifically from the 90 % to the 50 % case, which presented a BD-PSNR varying between 0 and 0.06 dB and a BD-rate increase between 0.03 and 1.28 %. On the other hand, when the target time is set to values from 40 % to 20 %, the computational complexity savings cause a BD-PSNR decrease between 0.18 and 0.94 dB and a BD-rate increase between 3.98 and 22.64 %.

These two operational regions are clearly defined in Fig. 5.20, which shows the effect of the computational complexity scalability on the encoder R-D performance. Just as in Table 5.10, the figure shows separate results for the different spatial resolutions. The first operational region, at the right side of the vertical dashed line, presents computational complexities ranging from 50 to 100 % and provides an encoding process in which the R-D performance is minimally affected by the complexity-constraining algorithm. The second region, at the left side of the vertical dashed line, allows further scaling of the computational complexity (down to 20 %) at the cost of higher R-D performance losses. Notice, however, that even though these losses are not negligible, they are acceptable in many applications and situations that require video encoding with very low computational complexity and tolerate some image quality decrease as a trade-off. Figure 5.20 also shows that the proposed method yields better R-D performance when scaling computational complexity of high-resolution video encoding.

The R-D performance for each target complexity tested is presented in Fig. 5.21 for the best and worst cases, which correspond to the *Traffic* (Fig. 5.21a) and *RaceHorses1* (Fig. 5.21b) sequences, respectively. In both cases, it is possible to

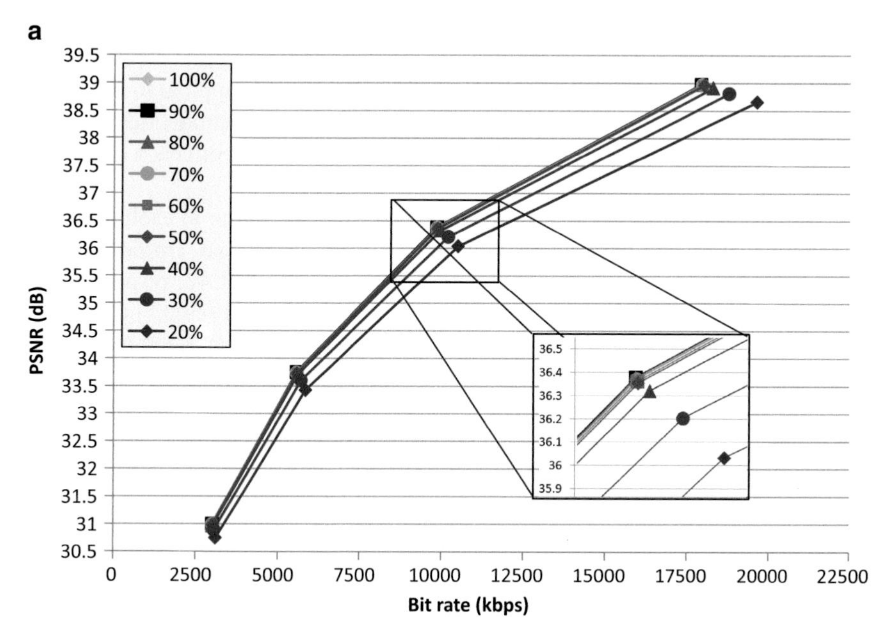

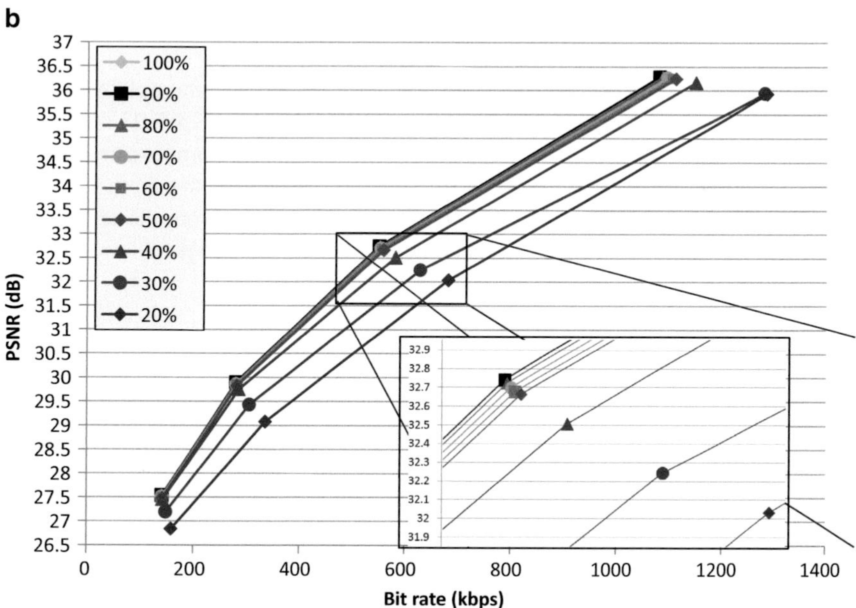

Fig. 5.21 R-D performance for each target complexity for videos (**a**) *Traffic* and (**b**) *RaceHorses1* encoded with QPs 27, 32, 37 and 42

notice that the difference between results for target complexities between 50 and 100 % is very small, especially in Fig. 5.21a. In fact, the R-D results are so similar that these curves are only visible individually when a zoom is applied to the chart, as shown in the dotted box in both charts. When smaller target complexities are used (20, 30 and 40 %), the corresponding curves are more distant from each other due to larger R-D performance losses.

Figure 5.22a–d presents the average encoding time per frame in each GOP for four video sequences (*PartyScene*, *vidyo1*, *BasketballDrive* and *Traffic*, respectively), each one encoded with a different QP (27, 32, 37 and 42, respectively). Besides the case in which no complexity scalability is applied (continuous black line at the top, labelled as "100 %"), the 20 % (red), 40 % (green), 60 % (blue) and 80 % (grey) cases are presented in the charts together with their respective target times (dashed lines in the same colour). The remaining target times tested are not shown in these figures for the sake of clarity, but they exhibit similar behaviour. As it can be seen, in all cases the actual encoding times vary a little around the target times from GOP to GOP. This is a normal effect caused by the variation of the video signal characteristics over time, which requires different encoding operations and different computational resources. Variations in encoding time can also be noticed when no complexity scalability is used (100 % case). The other video sequences presented behaviour similar to that shown in Fig. 5.22. The average difference between encoding time and target time for the four sequences presented in Fig. 5.22 is 4.6 %, 1.3 %, 2.3 % and 0.94 %, respectively.

The specific case of the *RaceHorses1* sequence, which was the only exception in the results presented in Table 5.10 and Fig. 5.19, is shown in Fig. 5.23 for QP 32. It is possible to perceive that even though the encoding times do not scale down to values as close to the target times as in the other sequences, the method is still capable of decreasing the encoding time to levels not too far from the target. The average difference between encoding and target time for such sequence was 15 %.

5.7 Conclusions

The contribution of this chapter consists in a set of algorithms for scaling the computational complexity of an HEVC encoder. The R-D efficiency as well as the complexity scaling accuracy of each method was separately presented and discussed in their corresponding sections. In this section, the five methods are compared and their results are jointly discussed. The average results presented in Tables 5.1, 5.2, 5.4, 5.6 and 5.11 for the FDCS, VDCS, MCTDL, CTDE and CCUPU methods, respectively, were all combined in the charts presented in Figs. 5.24, 5.25 and 5.26 in terms of complexity scaling accuracy, BD-rate increase and BD-PSNR decrease, respectively.

Figure 5.24 shows that, in general, the complexity scaling accuracy of the five methods is quite similar, since the resulting running complexity is always close to the ideal case (dotted grey line). The figure also shows that the computational

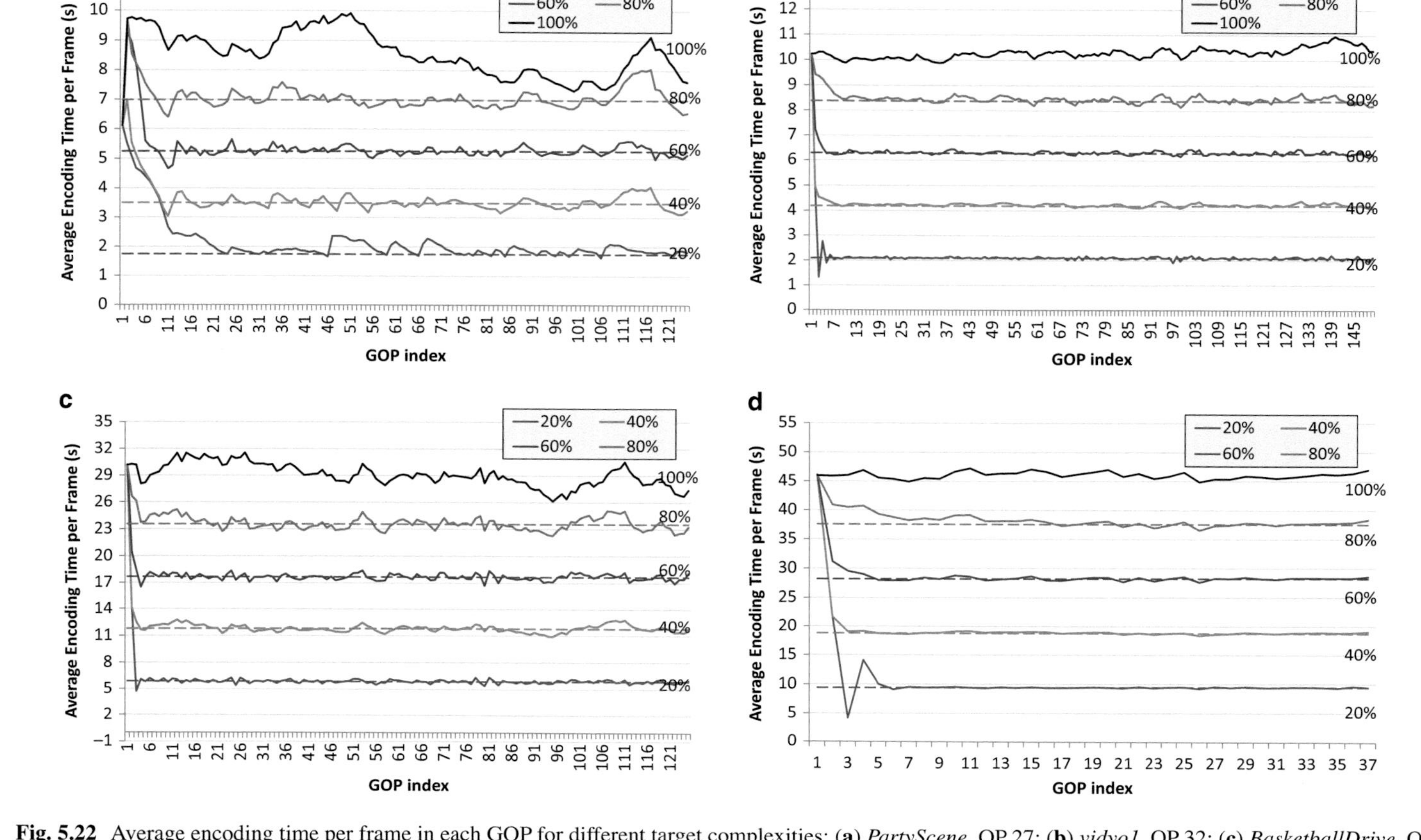

Fig. 5.22 Average encoding time per frame in each GOP for different target complexities: (**a**) *PartyScene*, QP 27; (**b**) *vidyo1*, QP 32; (**c**) *BasketballDrive*, QP 37; and (**d**) *Traffic*, QP 42

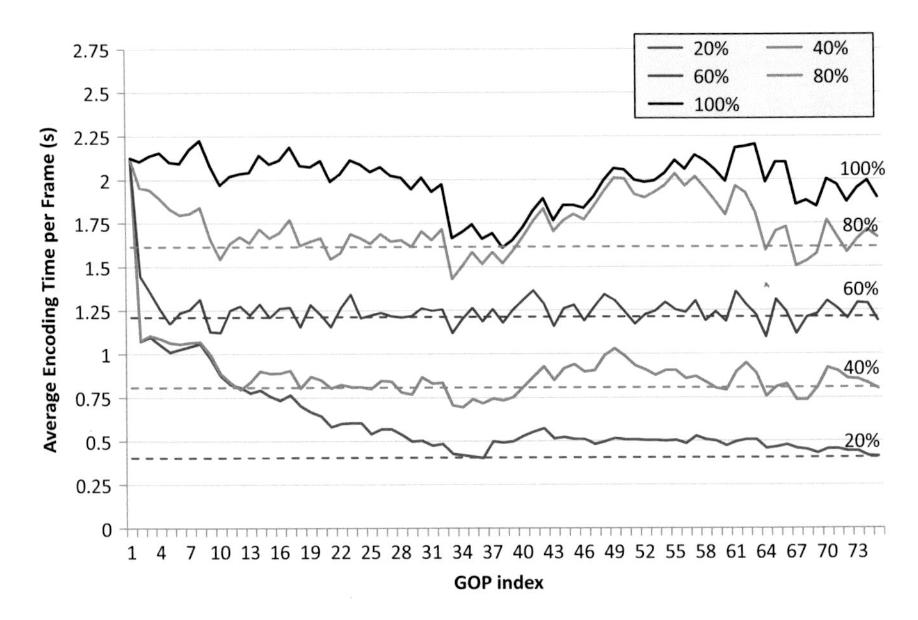

Fig. 5.23 Average encoding time per frame in each GOP for different target complexities, considering video *RaceHorses1* and QP 32

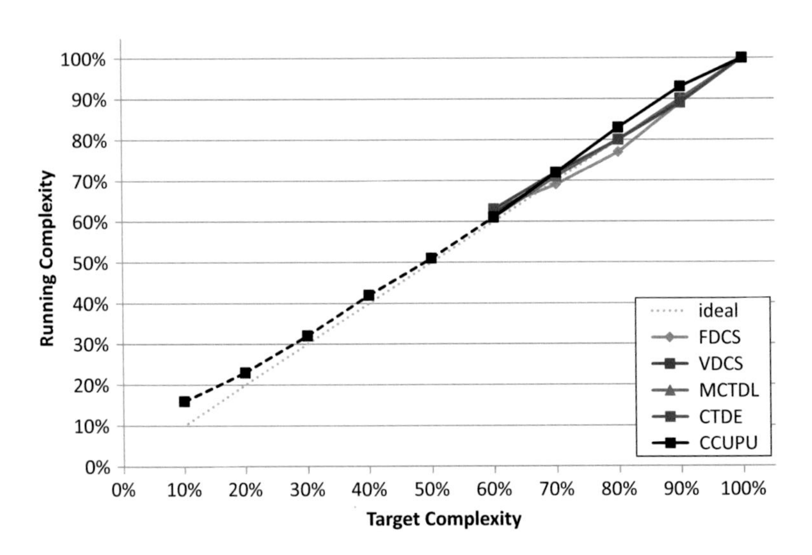

Fig. 5.24 Target complexity versus running complexity for the complexity scaling methods

complexity can be scaled down to targets lower than 60 % with the CCUPU method (dashed portion of the black curve), which is not possible with the remaining algorithms. The largest difference between target and running complexity appears, as already mentioned, when a target complexity of 10 % is set for the CCUPU method. In that case, the complexity achieved is around 16 %, which is still close to the target.

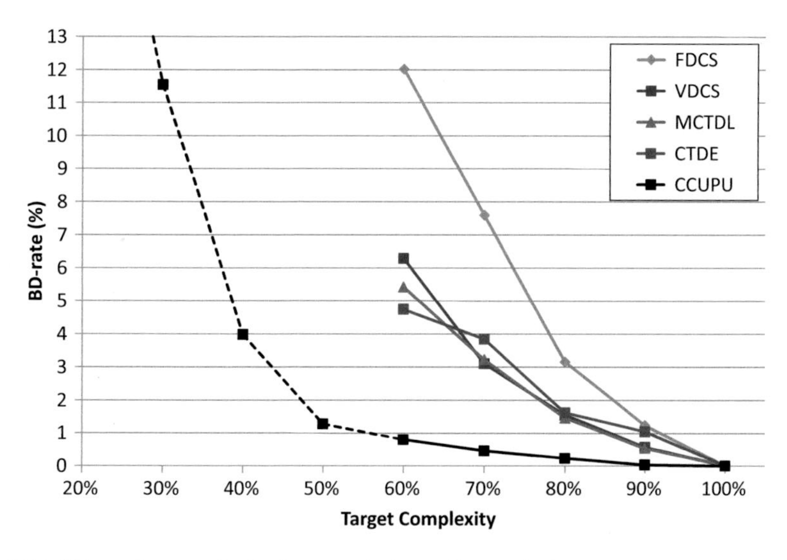

Fig. 5.25 BD-rate increase for the five complexity scaling methods

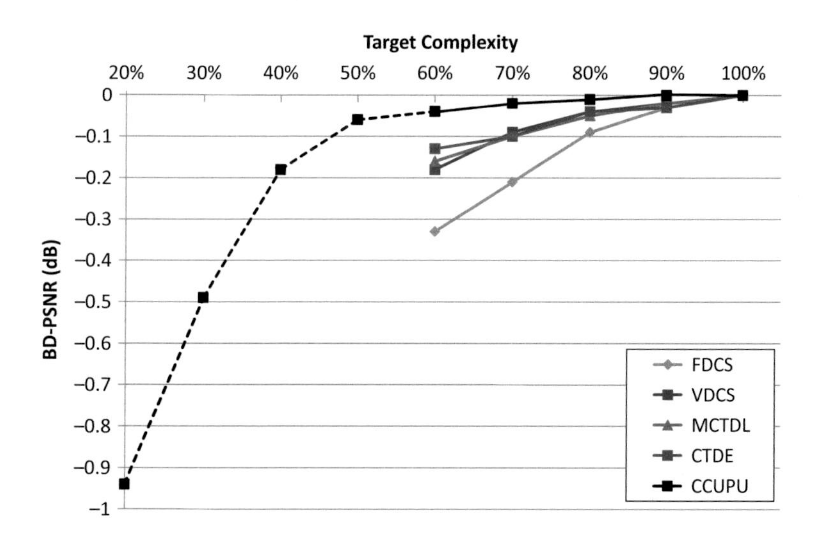

Fig. 5.26 BD-PSNR for the five complexity scaling methods

Figure 5.25 presents the BD-rate increase caused by applying different target complexities to the five methods. It is possible to notice from the first method developed (FDCS) to the last one (CCUPU) an increasing R-D efficiency, which is an outcome obtained by adding up more intelligent ways of constraining the frame partitioning structures of HEVC. In FDCS, the maximum coding tree depth was constrained for the whole frame, which is a rather crude way of scaling the computational complexity. In contrast, in CCUPU the encoding of each CTU is independently adjusted, and up to two levels of constraining can be performed per CTU. This reflects directly in the BD-rate and BD-PSNR results presented in Figs. 5.25 and 5.26. For example, when the target complexity is set to 50 %, the BD-rate increase caused by the CCUPU method is smaller than the BD-rate increase caused by FDCS with target complexity set to 90 %. Similarly, the CCUPU method is able to reduce complexity to 30 % at a smaller BD-rate cost than FDCS in 60 %.

The results compared in this section indicate that considering spatial and temporal correlation of video sequences in the complexity scaling algorithms (as done in VDCS, MCTDL and CTDE) decreases the R-D efficiency loss caused by FDCS. Furthermore, when constraining the PU splitting mode decision process before constraining the coding tree decision (as done in CCUPU), the computational complexity was further reduced at a much smaller R-D efficiency loss. These results show that the HEVC encoding complexity can be efficiently adjusted at small or negligible R-D losses if smart approaches are employed in the constraining process. If more information about the original video sequence characteristics, such as motion activity profile and texture complexity, as well as intermediate results obtained during the encoding process, were considered when adjusting the encoder complexity, R-D efficiency losses even smaller than those of CCUPU could be achieved. Chapter 6 furthers these ideas by presenting a new complexity reduction approach based on data mining and machine learning, which achieves even better results.

References

1. G. Correa, P. Assuncao, L.A. da Silva Cruz, L. Agostini, Dynamic tree-depth adjustment for low power HEVC encoders, in *Electronics, Circuits and Systems (ICECS), 2012 19th IEEE International Conference on* (2012), pp. 564–567
2. G. Correa, P. Assuncao, L.A. Da Silva Cruz, L. Agostini, Adaptive coding tree for complexity control of high efficiency video encoders, in *2012 Picture Coding Symposium* (2012), pp. 425–428
3. G. Correa, P. Assuncao, L. Agostini, L.A. da Silva Cruz, Complexity control of high efficiency video encoders for power-constrained devices. IEEE Trans Consumer Electron **57**, 1866–1874 (2011)
4. G. Correa, P. Assuncao, L. Agostini, L.A. da Silva Cruz, Motion compensated tree depth limitation for complexity control of HEVC encoding, in *2012 IEEE International Conference on Image Processing* (2012), pp. 217–220
5. G. Correa, P. Assuncao, L. Agostini, L.A. da Silva Cruz, Coding Tree Depth Estimation for Complexity Reduction of HEVC in *2013 Data Compression Conference*, Snowbird, UT (2013), pp. 43–52
6. G. Correa, P. Assuncao, L. Agostini, L.A. da Silva Cruz, Complexity control of HEVC through quadtree depth estimation, in *EUROCON, 2013 IEEE* (2013), pp. 81–86
7. G. Correa, P. Assuncao, L.A. Da Silva Cruz, L. Agostini, Constrained Encoding Structures for Computational Complexity Scalability in HEVC, in *2013 Picture Coding Symposium*, San Jose, USA (2013)
8. G. Correa, P. Assuncao, L. Agostini, L. Silva Cruz, Complexity scalability for real-time HEVC encoders. J Real-Time Image Proc. 1–16 (2014) http://link.springer.com/article/10.1007/s11554-013-0392-8
9. ISO/IEC-JCT1/SC29/WG11, Common test conditions and software reference configurations, Geneva, Switzerland (2012)

Chapter 6
Complexity Reduction for HEVC Using Data Mining Techniques

As the experiments presented in Chap. 4 have shown and the complexity scaling methods of Chap. 5 have confirmed, an important share of the high computational complexity of HEVC comes from the use of very flexible partitioning structures, such as the CUs, the PUs and the TUs. This chapter describes the process of using data mining (DM) techniques to build a set of models that are used to decide if the RDO-based partitioning structure decision process should be terminated early or run to its full extent [1–4]. By using information from intermediate encoding variables collected during the encoding of a set of video sequences, a set of decision trees were built and implemented in the HM encoder. When using this modified encoder, the operation of the decision trees sidesteps the encoder from having to run the full RDO process to find the best partitioning structures. The study of correlations and information gains associated with each variable, recorded while encoding test videos with the original HM encoder, was essential to the development of the early termination schemes presented in this chapter.

As explained in previous chapters, among the three main partitioning structures of HEVC, CUs have a central role due to their interdependence with the remaining partitioning structures. This means that by changing the number of coding tree configurations tested for a CTU, the overall number of tests performed to define PUs and RQTs is also affected. For this reason, the first early termination investigated in this chapter focuses on the coding tree determination process. The second early termination method proposed focuses on the PU splitting mode decision, since the experiments presented in Sect. 5.6.1 have shown that limiting the PU splitting mode decision results in significant computational complexity decreases with small compression efficiency losses. Finally, the third early termination focuses on the RQT decision, which is the partitioning structure that presents the smallest impacts on the encoding computational complexity.

When separately implemented, the early termination schemes achieved an average computational complexity reduction varying from 7.2 % up to 50 % for a negligible encoding performance reduction ranging from 0.05 % up to 0.56 % in terms of BD-rate increase. When jointly implemented, an average computational

G. Corrêa et al., *Complexity-Aware High Efficiency Video Coding*,
DOI 10.1007/978-3-319-25778-5_6

complexity reduction of up to 65% is achieved, with a small BD-rate increase of 1.36%. Extensive experiments and comparisons demonstrate that the proposed schemes achieve the best R-D-C trade-offs among all comparable works.

6.1 Introduction to Data Mining and Decision Trees

Knowledge Discovery from Data (KDD) is an interdisciplinary subfield of computer science currently applied to several areas, such as medicine, market management, biology and image processing. The goal of KDD systems is to extract information from both structured and unstructured sources by using DM and machine learning algorithms. In the work presented in this chapter, a predictive DM approach is used.

Predictive DM techniques are used to determine the value of dependent variables by looking at the value of some attributes in the data set, identifying regularities and building generalisation rules that can be expressed as models. There are several methods of predictive DM currently available, which vary broadly from one another in terms of efficiency, complexity and applicability. Decision trees [5] are a type of commonly used predictive DM, in which a dependent variable can assume one among a finite number of outcomes. In classification trees, which are a specific type of decision trees used in the work presented in this chapter, the dependent variable is called the *class* attribute, and it can take a finite number of outcomes.

When building decision trees, observations on a set of training data are mapped into arcs and nodes, as shown in the example given in Fig. 6.1. The inner nodes (A, B, C, D, in Fig. 6.1) represent the variables (attributes) tested, while the arcs are the possible values that the attributes can assume. In a binary classification tree, such as those designed and used in the context of this research, the attributes can assume two results ($x1$ and $y1$ for attribute A in the example of Fig. 6.1). Finally, the leaves of decision trees are the values that the class attribute can assume and

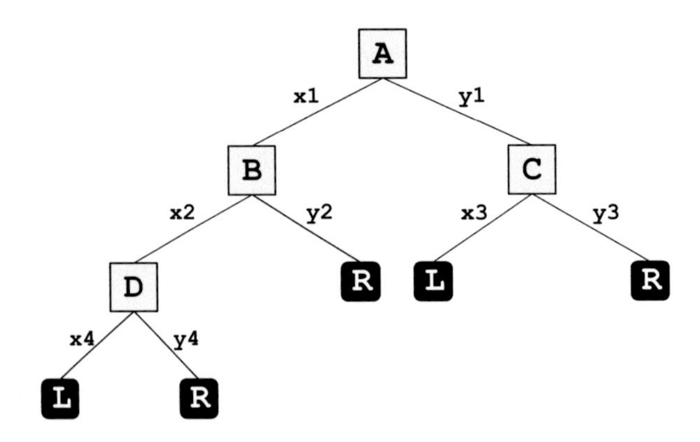

Fig. 6.1 Example of a binary classification tree

represent the possible outcomes of the whole decision process. In the example given in Fig. 6.1, the possible outcomes are L and R.

When the decision tree is implemented, in order to classify a determined instance into one of the possible outcomes, the algorithm starts at its root (A, in Fig. 6.1), tests the corresponding attribute value and descends to the next node through the appropriate branch, depending on the test result. For example, a data instance with feature vector ($A=x1$, $B=x2$, $C=y3$, $D=y4$) would be classified as R in the decision tree of Fig. 6.1.

Decision trees are commonly used mainly because of the following characteristics:

1. They usually achieve high prediction accuracy after trained.
2. They are easily understood by human beings and therefore simple to be implemented.
3. There are many efficient algorithms to build them from training data.
4. They can deal with both categorical and numerical values.
5. Once implemented, they execute predictions very fast.

The first and the fifth characteristics above are extremely important for the research work presented in this chapter, which aims at reducing the computational complexity of HEVC encoders without harming its R-D efficiency. This can only be performed if good prediction accuracy is achieved with the obtained decision trees and if the prediction process adds negligible extra computational complexity to the encoder.

6.2 Methodology

The *Waikato Environment for Knowledge Analysis* (WEKA) [6] was used to aid the DM process described in this chapter. WEKA is a free, open-source DM tool that includes several machine learning algorithms for preprocessing, classifying, clustering and visualising the data set, supporting statistical evaluation of learning schemes. Data fed to WEKA was obtained through offline encodings of the following ten video sequences, in which characteristics can be found in Appendix A: *BlowingBubbles*, *RaceHorses1*, *PartyScene*, *BQMall*, *SlideShow*, *vidyo1*, *BasketballDrive*, *ParkScene*, *NebutaFestival* and *Traffic*. All video sequences were encoded with QPs 22, 27, 32, 37 and 42, using the *Random Access* temporal configuration [7]. The HM software—version 12 (HM12)—was modified to save internal variables with intermediate encoding results into files that were used to create the training files used by WEKA.

The input for WEKA is ARFF (*Attribute-Relation File Format*) files, containing plain text describing a list of instances sharing a set of attributes. Figure 6.2 shows an ARFF file example, with its two major sections separated by dashed boxes. The first section consists of a header with the name of the relation and the attribute's declaration (i.e. name and type of values of that attribute). The second section

```
@RELATION CodingTreeEarlyTermination                          Header

@ATTRIBUTE RDcost_MSM NUMERIC
@ATTRIBUTE RDcost_2Nx2N NUMERIC
@ATTRIBUTE RDcost_2NxN NUMERIC
@ATTRIBUTE RDcost_Nx2N NUMERIC
@ATTRIBUTE part {0,1,2,3,4,5,6,7}
@ATTRIBUTE MergeFlag {0,1}
@ATTRIBUTE SkipMergeFlag {0,1}
@ATTRIBUTE neighDepth NUMERIC
@ATTRIBUTE SplitCU {0,1}
```

```
@DATA                                                         Raw Data

839  3214 2801 2801 0  1  1  0.25     1.75     0
3920 5055 4421 4421 0  0  0  2.13542  0.13541  1
990  2617 2687 2148 0  1  1  0.875    1.125    0
...
5505 5001 5307 4895 7  1  1  1.95833  0.041667 1
```

Fig. 6.2 Example of an ARFF file

contains the training data with one instance per line and one attribute value per column. In the specific case of building decision trees, the last line of the first section identifies the class attribute, for which the machine learning algorithms try to find a general (prediction) rule. In the example of Fig. 6.2, *SplitCU* is a class attribute that can assume a binary value (i.e. either 0 or 1). This is also the case of all decision trees proposed in this chapter.

To reduce the problem of the class data imbalance, which occurs when there are significantly more training instances belonging to one class than to the other(s), following common practice [8], the ARFF files used in the evaluation are composed of data sets with half the instances classified into each class (e.g. 50 % classified as 0 and 50 % classified as 1 in the example of Fig. 6.2). This was accomplished by resampling the training data instances when building the final ARFF files, which are composed of equal numbers of random samples of the data collected during the encoding of the ten video sequences encoded with the five QPs used.

For each early termination scheme proposed in this chapter, several variables were recorded during execution of the HM encoder, such as the sum and the variance of luminance samples in a CU, the absolute sum and variance of prediction residues in a PU, the horizontal and vertical gradients in a possible PU edge and the R-D cost of each PU splitting mode. The usefulness of each of these variables for the decision tree was assessed through the *Information Gain Attribute Evaluation* (IGAE) method in WEKA, which measures the information gain [9] achievable by using a variable to classify the data into the different classes. This gain equates to the difference between the number of bits per data item necessary to convey its class identity before and after classification of the data set using decision rules based on the variable in question [5]. Therefore, the information gain of a variable indicates how relevant it is for the process of constructing a decision tree that correctly decides to which class each data item belongs. In the case of the WEKA software, this information gain is measured by the *Kullback-Leibler divergence* (KLD) [10] of the pre- and post-classification probability distributions. Based on this measure (IGAE), a manual analysis procedure was followed to identify the most useful variables for the (tree) decision processes. Then, the variables with higher information gain were selected as attributes for the tree training processes, as explained later in this chapter.

The training of the decision trees was performed with the *C4.5* algorithm [5], which also uses KLD to choose the best attribute for each decision step and the thresholds corresponding to each decision step. The *C4.5* algorithm starts by taking all instances fed to it as inputs and calculates the information gain of using each attribute to perform the classification using a determined threshold. By iterating among all attributes and adjusting the thresholds, *C4.5* measures the information gain of each variable and threshold pair. Then, the attribute (and its corresponding threshold) with the largest information gain is chosen to divide the training data into two subsets. The same process is applied recursively to the two subsets. More details on *C4.5* can be found in [5].

The accuracy of all obtained trees was measured with WEKA by applying a tenfold cross-validation process. The level of accuracy was measured by the percentage of correct decisions in the total amount of instances used in the training process. Then, all trees obtained in the training phase were implemented in the HM software following a scheme of tests designed to allow a clear evaluation of each decision tree performance either individually or combined with others. A total of seven low-complexity schemes were implemented, and the encoding R-D efficiency was measured using sequences different from those used in the training phase. The seven schemes and their respective acronyms are:

1. Early termination for determining coding trees (*CT ET*) [3, 4]
2. Early termination for determining prediction units (*PU ET*) [1, 2, 4]
3. Early termination for determining residual quadtrees (*RQT ET*)
4. Joint early terminations for determining coding trees and prediction units (*CT+PU ET*) [4]
5. Joint early terminations for determining prediction units and residual quadtrees (*PU+RQT ET*) [4]
6. Joint early terminations for determining coding trees and residual quadtrees (*CT+RQT ET*) [4]
7. Joint early terminations for determining coding trees, prediction units and residual quadtrees (*CT+PU+RQT ET*) [4]

6.3 Early Termination for Determining Coding Trees

The proposed coding tree early termination [3, 4] consists in deciding whether or not the splitting of CUs into four smaller CUs should be tested. If the decision tree outcome is *yes* (i.e. $SplitCU=1$), the current CU is split into four smaller CUs and

Table 6.1 Information gain attribute evaluation for the coding tree early termination

	Information gain		
Attribute	64×64	32×32	16×16
Partition	0.352	0.336	0.269
ΔNeighDepth	0.311	0.262	0.249
Ratio(2N×2N, MSM)	0.112	0.168	0.255
NormDiffRD(2N×2N, MSM)	0.109	0.163	0.249
RD(2N×2N)	0.035	0.042	0.053
RD(MSM)	0.034	0.061	0.108
RD(2N×N)	0.033	0.036	0.044
RD(N×2N)	0.031	0.032	0.042
SkipMergeFlag	0.046	0.066	0.065
MergeFlag	0.020	0.035	0.046

the next coding tree depth is tested. Otherwise (i.e. *SplitCU*=0), the coding tree splitting process is halted in the current CU.

As the HEVC standard allows up to four coding tree depths, three different decision trees were created for the cases that allow splitting into smaller CUs: 64×64, 32×32 and 16×16. Table 6.1 shows the HM variables that provided best performance as measured by the information gain with respect to the decision of splitting or not splitting a CU (i.e. those that were selected as attributes for the decision trees). Information gain values are presented separately for each CU size in the table, and the attributes are described in detail in the following paragraphs.

The *Partition* attribute corresponds to which PU splitting mode was chosen for the current CU (i.e. $2N\times 2N$, $2N\times N$, $N\times 2N$, $N\times N$, $2N\times nU$, $2N\times nD$, $nL\times 2N$ or $nR\times 2N$), independently of whether inter- or intra-frame prediction was applied. The idea behind saving this information is that when a large PU (e.g. $2N\times 2N$) is chosen as the best option to predict a determined CU, further tests to determine the coding tree configuration are probably not necessary, so that this CU does not need being split into smaller sub-CUs. Statistics that support this claim are presented in Fig. 6.3. The chart shows that most of the CUs predicted as a $2N\times 2N$ PU were not split into sub-CUs. For example, 83 % of 64×64 CUs predicted as a $2N\times 2N$ PU did not need being split into four 32×32 CUs, as the leftmost black bar of Fig. 6.3a shows. Conversely, an average of 83.3 % of 64×64 CUs encoded with the remaining modes were split into four 32×32 CUs (average of all grey bars of Fig. 6.3a, except for the $2N\times 2N$ case). By analysing the three charts of Fig. 6.3, it is possible to notice that, on the one hand, the correlation between using $2N\times 2N$ PUs and not splitting the CU decreases in smaller CUs (70 % for 16×16 CUs, as shows Fig. 6.3c). On the other hand, in smaller CUs the correlation between choosing the remaining PU modes and splitting the CU increases (on average, 90 % for 16×16 CUs).

The *ΔNeighDepth* attribute is computed based on the difference between the coding tree depths used in neighbouring CTUs and the depth of the current CU. The rationale of considering such variable is that there exists a correlation among maximum depths of spatially and temporally neighbouring CTUs, as previously shown

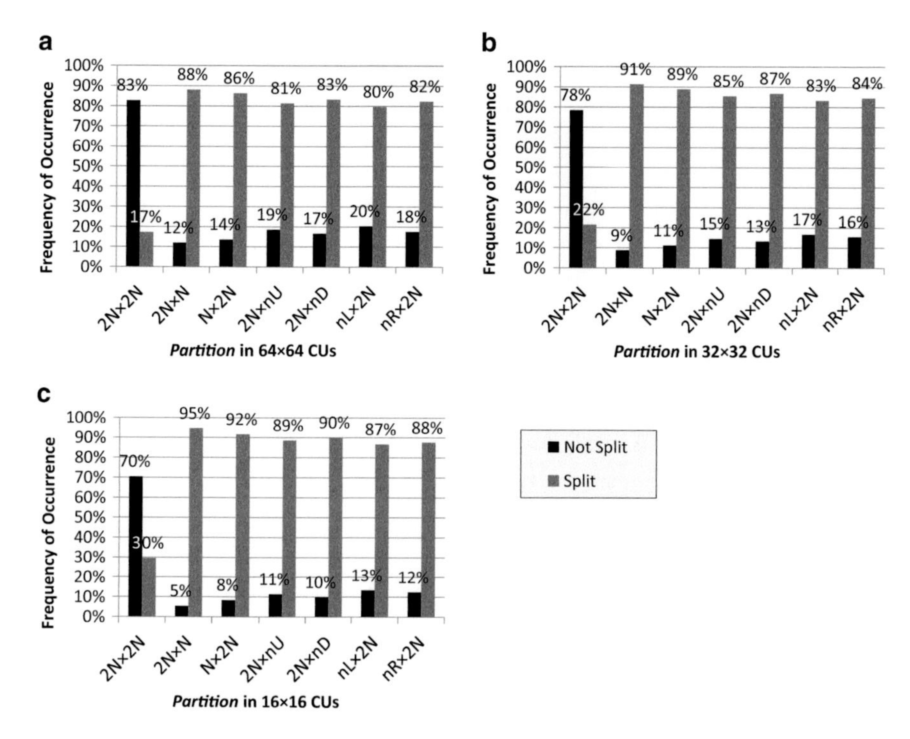

Fig. 6.3 Occurrence of (**a**) 64×64 CUs, (**b**) 32×32 CUs and (**c**) 16×16 CUs split and not split into smaller CUs according to the PU mode chosen (i.e. *Partition*)

in Chap. 5 (see Table 5.9). The attribute Δ*NeighDepth* is calculated as follows. First, for each neighbouring CTU, the average depth among all its composing CUs is computed. The top-left, top, top-right and left CTUs in the current frame, as well as the co-localised CTUs in the first frames of both reference lists (*List 0* and *List 1*), are considered as neighbours, so that up to six average depths are calculated. Figure 6.4 shows the four neighbouring CTUs in the current frame and the co-localised CTUs in the reference frames, with the variables representing the average CU depths assigned to them. Let us call these averages as A_1 to A_N, where N is the number of neighbouring CTUs available for the current CTU. The CTU assigned with label C in Fig. 6.4 represents the current CTU. Finally, the value of Δ*NeighDepth* is calculated as an average of the averages A_1 to A_N, minus the depth of the CU currently being encoded, as shown in Eq. (6.1). If the current CU depth is much smaller than the average of average depths of neighbouring CTUs, the splitting process should probably continue due to spatio-temporal correlation among neighbouring CTUs.

$$\text{Neigh Depth} = \frac{\sum_{i=1}^{N} A_i}{N} - \text{Curr Depth} \tag{6.1}$$

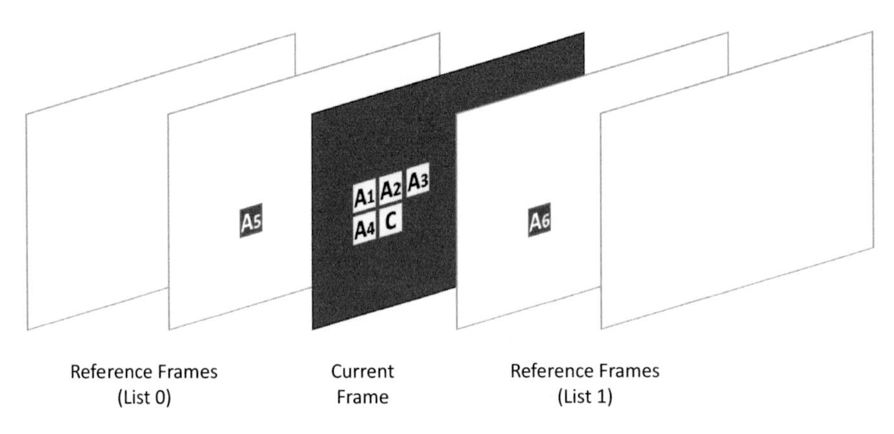

Fig. 6.4 Neighbouring CTUs used in the calculation of $\Delta NeighDepth$

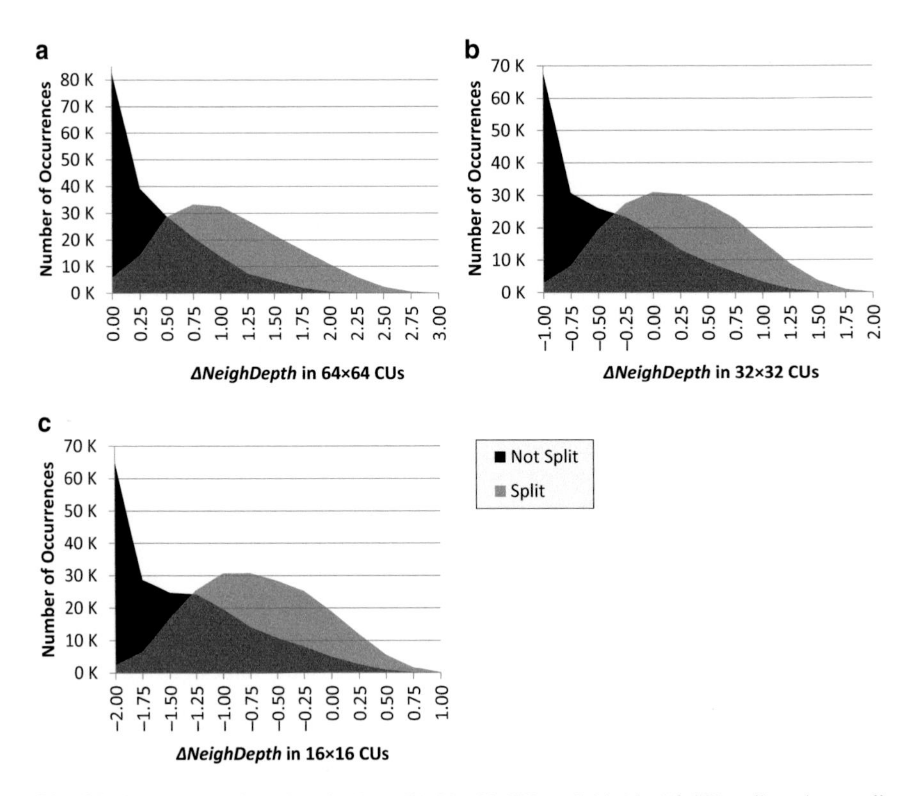

Fig. 6.5 Occurrence of (**a**) 64 × 64 CUs, (**b**) 32 × 32 CUs and (**c**) 16 × 16 CUs split and not split into smaller CUs according to the average of CU depths in neighbouring CTUs

Figure 6.5 shows the distribution of $\Delta NeighDepth$ for different CU sizes. The curves show that there is a clear relationship between the distribution of $\Delta NeighDepth$ and the CU splitting decision. CUs that are not split into smaller CUs have $\Delta NeighDepth$ values that cluster towards low magnitudes, while the opposite occurs for those CUs that are split into smaller CUs. Since the two distributions do not fully overlap, it is possible to determine an optimal decision threshold that minimises the classification error rate. WEKA computes these thresholds during the process of training the decision trees.

The *Ratio*($2N\times2N$, *MSM*) shown in Table 6.1 is calculated as a simple division between the R-D costs of encoding the current CU as an inter-predicted $2N\times2N$ PU and as an MSM PU, as shown in Eq. (6.2). The *NormDiffRD*($2N\times2N$, *MSM*) value is the normalised difference between the *RD*($2N\times2N$) and *RD*(*MSM*) costs, calculated as per Eq. (6.3). The reason for considering these values in the IGAE analysis is that when a compression gain (i.e. a drop in R-D cost) is observed due to the use of motion-compensated prediction in a CU instead of encoding it with MSM, the block probably belongs to a medium-/high-motion or complex-textured image region, and usually in this type of situation, it is advisable to split a CU into smaller CUs. Figure 6.6 shows the distribution of *Ratio*($2N\times2N$, *MSM*) for different CU sizes. The smaller information gain level of this parameter in comparison to the two previously analysed cases is also clear in the charts, which shows the *split* and *not split* areas more overlapped than in the charts of Figs. 6.3 and 6.5.

$$Ratio\left(2N\times2N, MSM\right)=\frac{RD\left(2N\times2N\right)}{RD\left(MSM\right)} \tag{6.2}$$

$$NormDiffRD\left(2N\times2N, MSM\right)=\left|\frac{RD\left(2N\times2N\right)-RD\left(MSM\right)}{RD\left(MSM\right)}\right| \tag{6.3}$$

The decision trees were trained with the attributes shown in Table 6.1 and their most important characteristics are detailed in Table 6.2. The table presents the decision accuracy of each tree, as well as their depth (i.e. number of sequential tests), number of test nodes and number of leaves. As Table 6.2 shows, the three obtained trees achieve a decision accuracy slightly above 84 %. The accuracies are measured by the ratio of the number of splitting decisions which agree with the splitting decision that would have been taken by the unmodified coder (both *Split* and *Not Split*) and the total number of CUs analysed in each case. However, these results count the case of splitting a CU that should not be split into smaller CUs as a decision error, even though it does not harm the encoding R-D efficiency, having as only deleterious effect an increase of the encoding complexity. In fact, the R-D efficiency is negatively affected by decision errors only when a CU that should be split is not split due to the early termination provided by the decision trees and so the CU is not as efficiently coded as it could be. The sixth column of Table 6.2 shows the percentage of such incorrect early terminations that could actually cause R-D efficiency losses due to inaccurate coding tree depth decisions. Regarding the topological characteristics of the decision trees, it is important to notice that all of them are composed of less than ten decision levels (*Depth* column in Table 6.2), which means that the computational complexity added to the encoder due to implementing such trees is negligible. Detailed descriptions of the trees obtained after training are available in the Appendix B of this book.

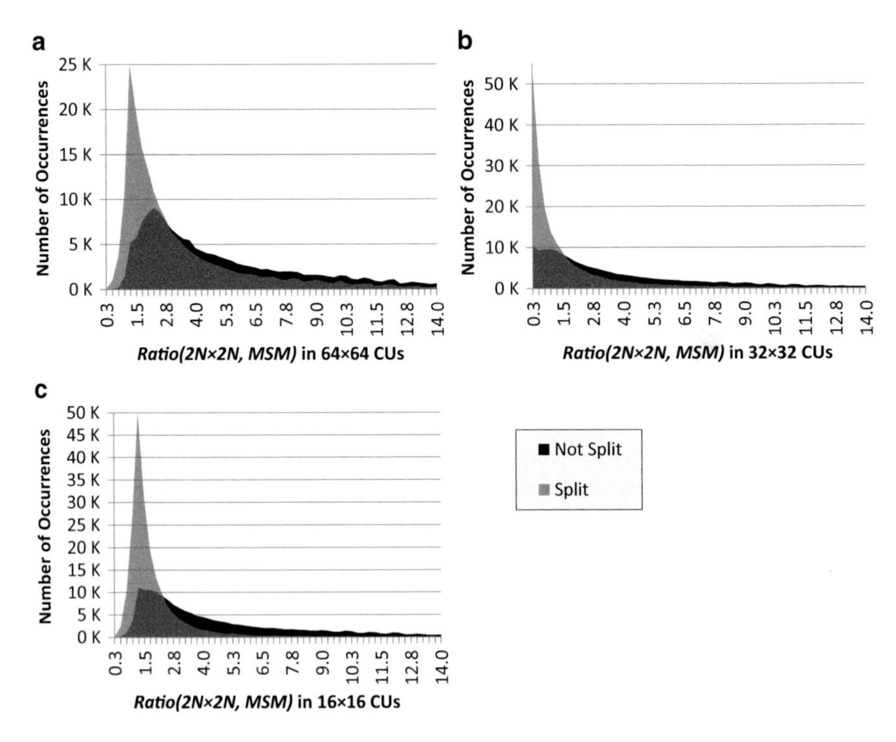

Fig. 6.6 Occurrence of (**a**) 64×64 CUs, (**b**) 32×32 CUs and (**c**) 16×16 CUs split and not split into smaller CUs according to the ratio between the $2N \times 2N$ and MSM R-D costs

Table 6.2 Characteristics and performance of trees trained for the coding tree early termination

CU size	Depth	Test nodes	Leaves	Decision accuracy (%)	Inaccurate depth (%)
64×64	5	6	19	84.2	7.1
32×32	8	20	33	84.5	7.5
16×16	9	23	44	84.6	6.9

Finally, to illustrate the effectiveness of the proposed method, Fig. 6.7 presents the 100th frame of the *BasketballDrill* video sequence and its corresponding CU boundaries according to the coding tree defined for each CTU. The sequence was encoded with QP 32 and the *Random Access* temporal configuration. Notice that the *BasketballDrill* video was not used in the training of the decision trees. The frame in Fig. 6.7a was encoded using the original HM encoder, while the frame in Fig. 6.7b was encoded with an HM encoder modified to include the coding tree early termination algorithm. It is possible to perceive that the differences between the boundaries in Fig. 6.7a, b are not expressive, which confirms that the early termination performs a correct coding tree determination in most cases.

a

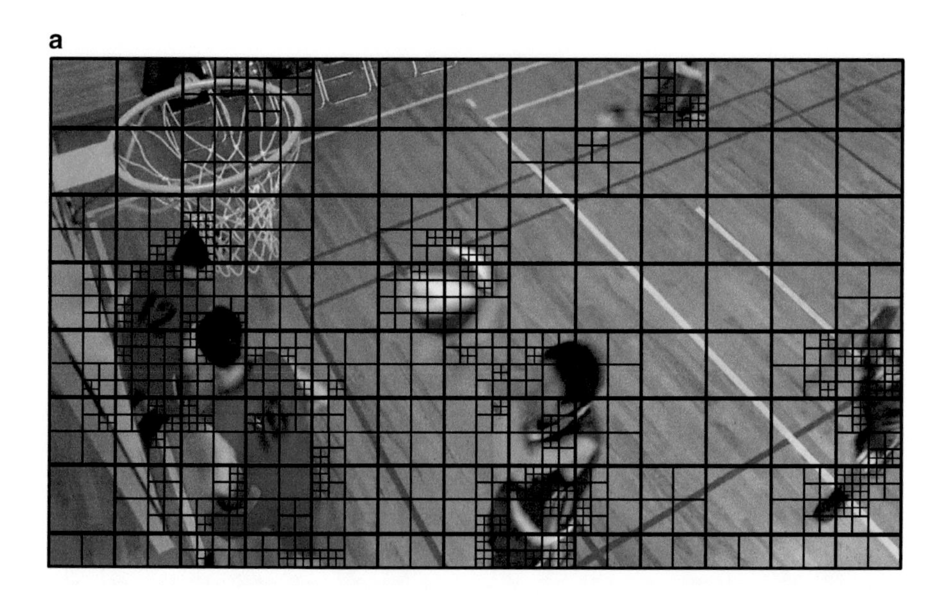

b

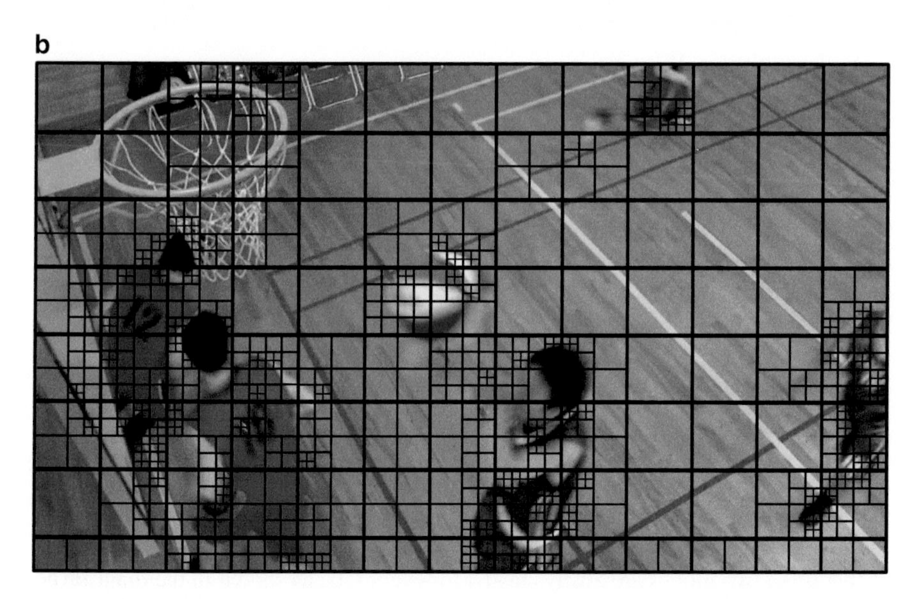

Fig. 6.7 CTUs divided into CUs in the 100th frame of the *BasketballDrill* video sequence encoded with QP 32 by (**a**) the original HM encoder and the (**b**) HM encoder with the coding tree early termination

6.4 Early Termination for Determining Prediction Units

The second early termination proposed [1, 2, 4] is motivated by the statistics presented in Fig. 6.8, which shows the average occurrence probability of each PU splitting mode in inter-predicted CUs. The statistics in the charts are for the

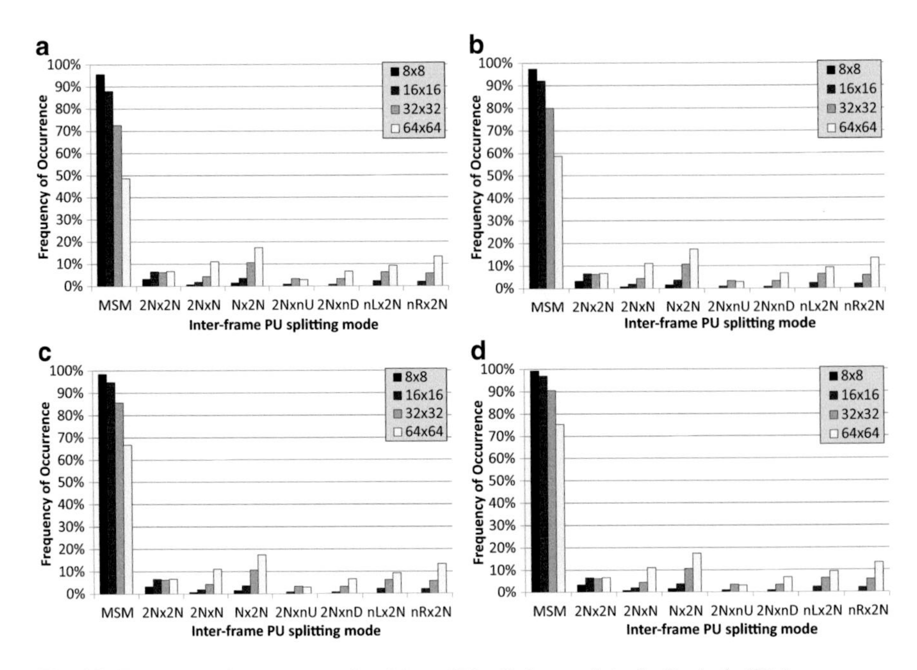

Fig. 6.8 Frequency of occurrence of each inter-PU splitting mode in the *BasketballDrive* sequence encoded with (**a**) QP 27, (**b**) QP 32, (**c**) QP 37 and (**d**) QP 42

BasketballDrive video sequence encoded with QPs 27, 32, 37 and 42. The remaining sequences presented similar behaviour. In Fig. 6.8, it is clear that most inter-predicted CUs are encoded without being split into smaller PUs and employ mainly the MSM mode. On average, 58, 76, 89 and 95 % of 64×64, 32×32, 16×16 and 8×8 CUs, respectively, are encoded with the MSM mode. However, even though the remaining modes are not so frequently used as MSM, especially in small CUs, they are still always tested in a full RDO-based decision, which is not ideal when trading off compression efficiency and computational complexity.

As in the case of early termination for coding trees presented in Sect. 6.3, the decision trees for early terminating the PU splitting mode decision were designed to decide whether or not the search for the best PU structure should continue after some PU splitting modes have been tested. As most inter-predicted CUs are encoded as a single PU, the decision trees are used after testing the MSM and $2N\times2N$ modes, so that these two modes are always tested for every CU, as shown in the diagram of Fig. 6.9. In case of early termination (decision labelled as *ET* in Fig. 6.9), the mode with smallest R-D cost between MSM and $2N\times2N$ is chosen for the CU. Otherwise (decision labelled as *keep* in Fig. 6.9), the remaining modes are tested.

As in the case presented in Sect. 6.3, 50 % of the training data come from inter-predicted CUs that have been split into PUs smaller than $2N\times2N$ and 50 % come from inter-predicted CUs that have not been split into PUs smaller than $2N\times2N$. Four different decision trees were built, each one for a different CU size (64×64, 32×32, 16×16 and 8×8).

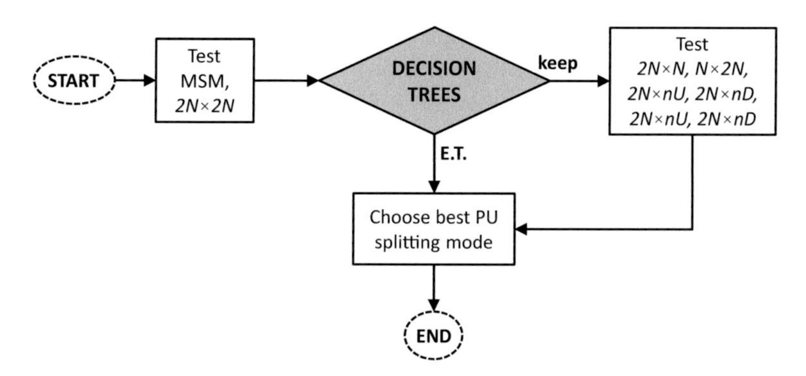

Fig. 6.9 Inter-frame PU splitting mode decision with the proposed early termination

Table 6.3 Information gain attribute evaluation for the PU early termination

	Information gain			
Attribute	64×64	32×32	16×16	8×8
Ratio(2N×2N, MSM)	0.245	0.390	0.475	0.572
NormDiffRD(2N×2N, MSM)	0.129	0.245	0.341	0.433
RD(MSM)	0.224	0.306	0.383	0.422
RD(2N×2N)	0.165	0.139	0.101	0.135
RD(best)	0.208	0.284	0.364	0.407
UpperCU_div	–	0.223	0.297	0.240
Ratio(best, MSM)	0.195	0.266	0.229	0.208
NormDiffRD(best, MSM)	0.146	0.207	0.203	0.186

In this early termination, the class attribute is the information of whether or not a CU should be split into PUs smaller than $2N\times 2N$. After an extensive observation on the collected data, the variables that provided the largest information gains were used for training the decision trees. These attributes are listed in Table 6.3, where information gain results are presented separately for each CU size.

The attributes *Ratio*($2N\times 2N$, *MSM*), *NormDiffRD*($2N\times 2N$, *MSM*), *RD*(*MSM*) and *RD*($2N\times 2N$) have the same meaning as in Sect. 6.3. Attribute *RD*(*best*) is the lowest R-D cost among the MSM and the $2N\times 2N$ modes, and *UpperCU_div* is the information of whether or not the CU in the upper coding tree depth was split into PUs smaller than $2N\times 2N$. Finally, the attributes *Ratio*(*best, MSM*) and *NormDiffRD*(*best, MSM*) correspond to the ratio and normalised difference (computed as in Eqs. (6.4) and (6.5), respectively), between *RD*(*best*) and *RD*($2N\times 2N$):

$$Ratio\left(best, MSM\right) = \frac{RD\left(best\right)}{RD\left(MSM\right)} \tag{6.4}$$

$$NormDiffRD\left(best, MSM\right) = \left|\frac{RD\left(\text{best}\right) - RD\left(MSM\right)}{RD\left(MSM\right)}\right| \tag{6.5}$$

The attribute that yields the highest information gain, on average, is the *Ratio*($2N\times2N$, *MSM*). An explanation similar to the one exposed in the previous section applies to justify the use of such attribute: when a compression gain is noticed due to performing ME/MC for a CU instead of encoding it with MSM, there is a chance of the block belonging to an image region with some motion activity or texture heterogeneity. In such cases, it is advisable to test smaller PU sizes to verify if they yield additional compression gains. Figure 6.10 presents statistical results in the form of distribution of the values of this attribute considering the four CU sizes. The statistics correspond to values obtained from all video sequences and QPs mentioned in Sect. 6.2. Figure 6.10 shows that the ratio between the $2N\times2N$ and MSM R-D costs is a relevant indicator of the necessity of testing the remaining modes. It is possible to see that most CUs with a small ratio were split into PUs smaller than $2N\times2N$, while CUs with larger ratio values are mostly encoded with $2N\times2N$ or MSM.

The relevance of the *UpperCU_div* attribute is illustrated in Fig. 6.11, which shows three charts for the attribute considering 32×32, 16×16 and 8×8 CUs. As 64×64 CUs do not have a parent CU, since they are the largest CUs allowed in HEVC, the *UpperCU_div* attribute is not used in their corresponding decision trees.

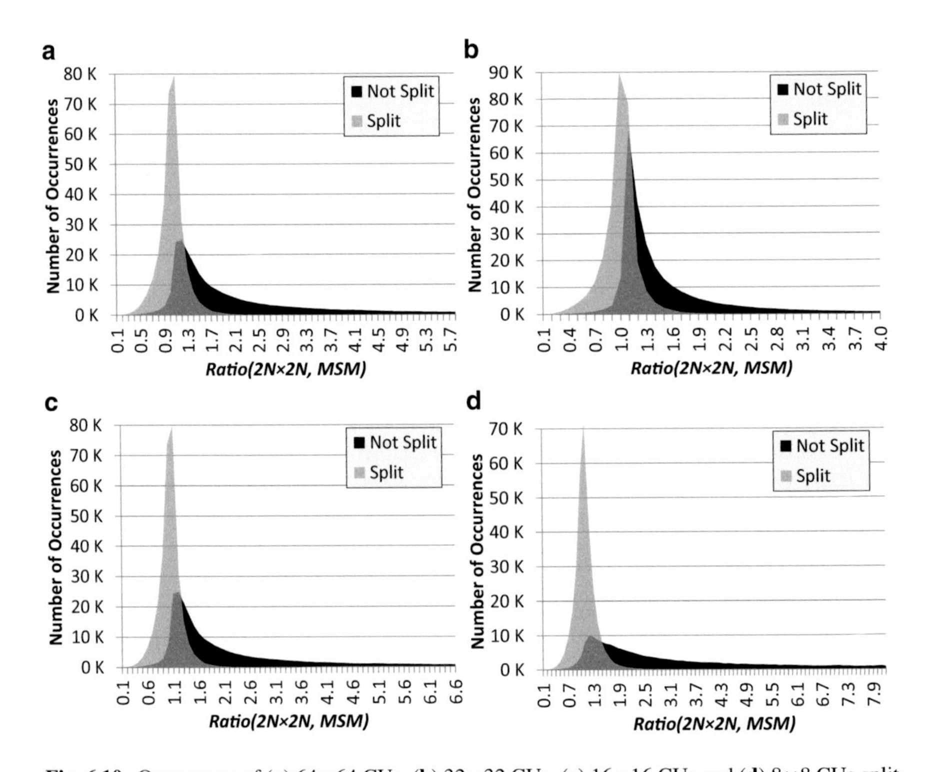

Fig. 6.10 Occurrence of (**a**) 64×64 CUs, (**b**) 32×32 CUs, (**c**) 16×16 CUs and (**d**) 8×8 CUs split and not split into PUs smaller than $2N\times2N$ according to the ratio between the R-D costs of $2N\times2N$ and MSM modes

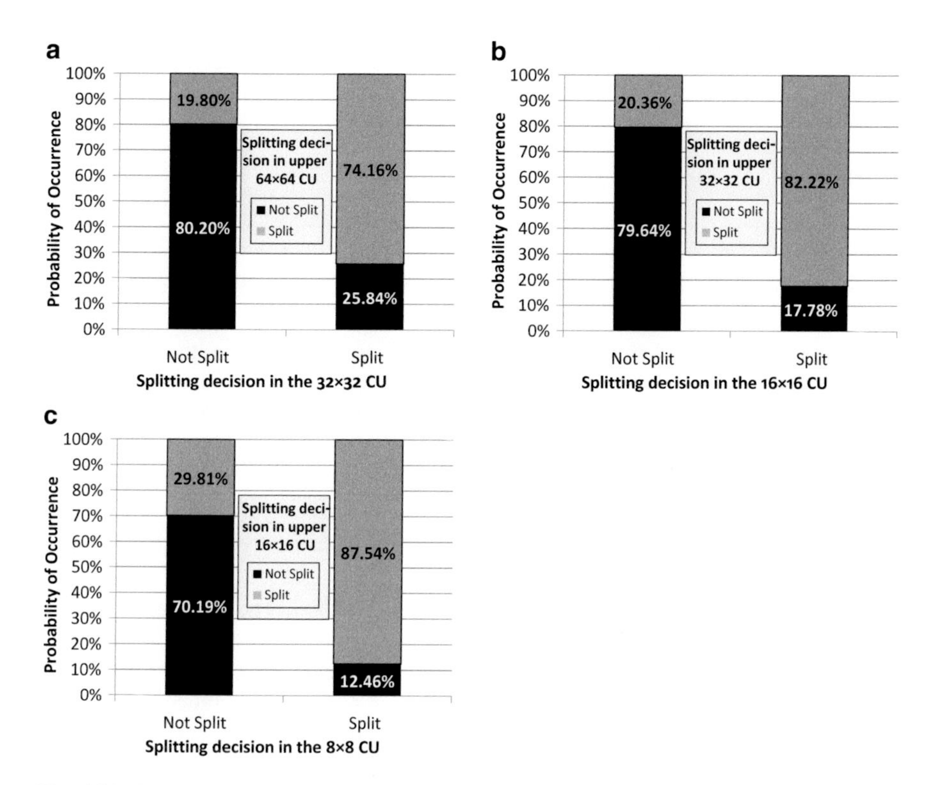

Fig. 6.11 Occurrence of (**a**) 32×32 CUs, (**b**) 16×16 CUs and (**c**) 8×8 CUs split and not split into PUs smaller than $2N \times 2N$ according to the splitting decision at the upper coding tree depth

The charts show that there is a strong correlation between the PU splitting in the upper and current CU depths. For example, Fig. 6.11b shows that in 82.22 % of the cases when a 16×16 CU was split into smaller PUs, its upper 32×32 CU was also split. Analogously, in 79.64 % of the cases when a 16×16 CU was not split into smaller PUs, its corresponding parent 32×32 CU was also not split. The charts also show that the correlation increases for the *Split* case in small CUs (8×8) and increases for the *Not Split* case in large CUs (32×32).

The decision trees were trained with the attributes shown in Table 6.3 and their characteristics are detailed in Table 6.4, which shows that their decision accuracy varies between 79.6 and 91.2 %. Accuracy values are measured by dividing the number of correct splitting decisions by the total number of CUs analysed. However, regarding inaccurate decisions, it is important to notice that R-D efficiency losses would occur only when a CU should be predicted with PUs smaller than $2N \times 2N$, but the decision process is early terminated, leading a nonoptimal PU splitting mode to be chosen. In the remaining inaccurate decisions (i.e. when the CU should be predicted with MSM or $2N \times 2N$ and the decision process is not early terminated), the encoder still chooses an optimal mode through RDO, since all modes are tested.

Table 6.4 Characteristics and performance of trees trained for the prediction unit early termination

CU size	Depth	Test nodes	Leaves	Decision accuracy (%)	Inaccurate depth (%)
64×64	5	8	9	79.6	8.6
32×32	7	9	10	86.0	5.0
16×16	5	9	10	89.2	4.0
8×8	4	5	6	91.2	2.1

The sixth column of Table 6.4 shows statistics only for the case that incurs in R-D losses, which varies between 2.1 and 8.6 %.

Table 6.4 also shows that all the trained trees are very short, with a maximum depth of 7 in the case of 32×32 CUs. This means that the complexity added to the encoder due to implementing such trees is negligible, especially when considering that all attributes are already calculated by the HM encoder during its normal operation, except for the R-D cost ratios and normalised differences, which calculations represent an insignificant computational overhead in comparison to the whole encoding process. Detailed descriptions of the four decision trees are presented in Appendix B.

To illustrate the effectiveness of the proposed method, Fig. 6.12 presents the 100th frame of the *BasketballDrill* video sequence, which was not used in the training of the decision trees, and its corresponding PU boundaries for each CU. The sequence was encoded with QP 32 and the *Random Access* temporal configuration. The frame shown in Fig. 6.12a was encoded by the original HM encoder, while the frame in Fig. 6.12b was encoded by the HM encoder with the PU early termination implemented. It is possible to perceive that there are few differences between the boundaries in Fig. 6.12a, b, which confirms that the early termination is capable of performing a correct PU determination in most cases.

6.5 Early Termination for Determining the RQT

As previously shown in Chap. 4, even though restricting the maximum TU size does not provide substantial computational complexity reductions, it only affects the encoding R-D efficiency marginally. If these restrictions are carefully performed by a trained early termination scheme, R-D efficiency losses would most probably be insignificant and some complexity reduction could still be achieved. For this reason, the last early termination presented in this chapter focuses on halting the process for determining the best RQT structure, as described in this section.

Similar to the coding tree early termination scheme (Sect. 6.3), the RQT early termination consists in deciding whether or not the splitting of TUs into four smaller TUs should be tried. If the tests return *yes*, the current TU is divided into four smaller TUs and the next RQT depth is tested. Otherwise, the process is finished for the current TU and its sub-TUs are not considered in the RDO-based decision.

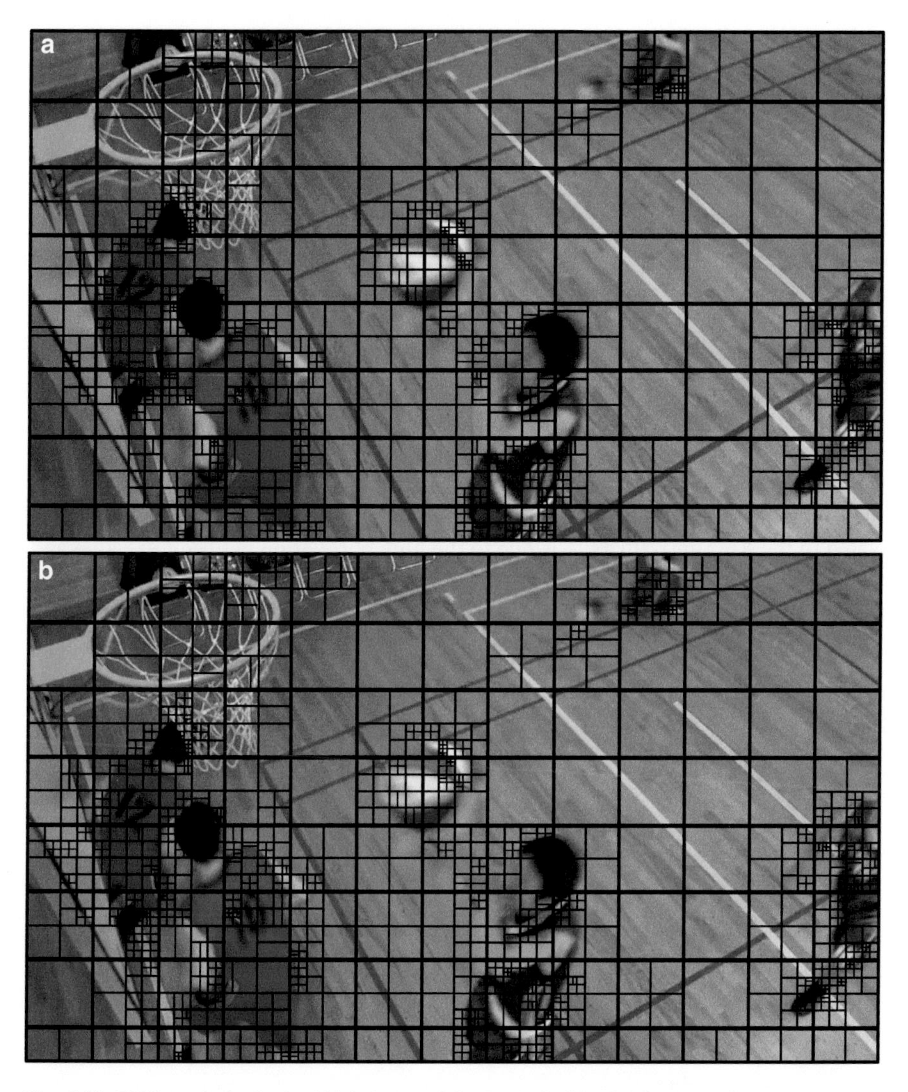

Fig. 6.12 PU boundaries in the 100th frame of the *BasketballDrill* video sequence encoded with QP 32 by (**a**) the original HM encoder and the (**b**) HM encoder with the PU early termination

As in the cases of the previously presented early terminations, the ARFF files used in the training process are 50 % composed of data from TUs that have been split into four smaller TUs and 50 % from TUs that have not been split. Given that TUs can assume three different dimensions in the *Main* encoder configuration used in the context of this book (32×32, 16×16 and 8×8), two of which can be split into smaller TUs (32×32 and 16×16), only two different decision trees were designed for this early termination scheme.

In each training data set, the class attribute is the information of whether or not a TU should be split into four TUs. The remaining variables obtained from HM were analysed with IGAE, and those selected as attributes to be used in the training of the decision trees are presented in Table 6.5, with their respective results in terms of information gain.

In Table 6.5, the attributes *AbsSumY*, *AbsSumU* and *AbsSumV* are the sum of absolute values from the luminance, blue chrominance and red chrominance residues, respectively, and the *nonZeroCoeffY*, *nonZeroCoeffU* and *nonZeroCoeffV* attributes represent the number of nonzero coefficients obtained after transforming the TU as a whole block (i.e. before splitting it). Finally, the *SingleCost* attribute is the R-D cost of encoding a TU as a whole block instead of splitting it.

As the residue samples are the inputs to the transform modules, we have investigated their probability of being encoded with small-sized TUs according to the intensity of its values. The values of *AbsSumY*, *AbsSumU* and *AbsSumV* attributes were analysed in order to check if the number of split TUs tend to increase with the sum of the residue samples. Figure 6.13 shows that the high information gain associated to the *AbsSumY* attribute is due to the high correlation between its value and the splitting decision when the absolute sum of residues is null. Figure 6.13a shows that 82.24 % of the 32×32 TUs with *AbsSumY* value equal to zero were not split

Table 6.5 Information gain attribute evaluation for the RQT early termination

Attribute	Information gain	
	32×32	16×16
AbsSumY	0.342	0.279
AbsSumU	0.057	0.033
AbsSumV	0.055	0.031
SingleCost	0.145	0.140
nonZeroCoeffY	0.348	0.284
nonZeroCoeffU	0.342	0.280
nonZeroCoeffV	0.342	0.280

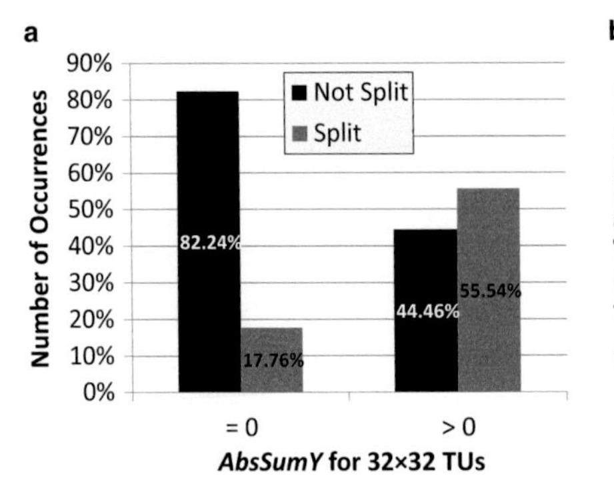

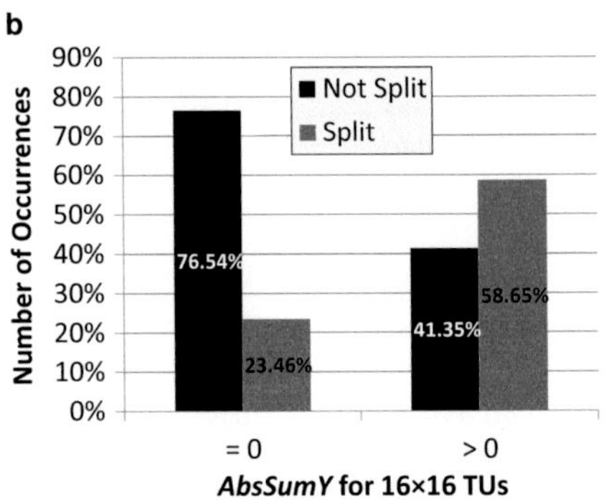

Fig. 6.13 Occurrence of (**a**) 32×32 and (**b**) 16×16 TUs split and not split into smaller TUs according to the absolute sum of luminance residues

into smaller TUs. The charts also show that this correlation decreases significantly when *AbsSumY* is larger than zero. Notice, however, that in typical video sequences, the value of *AbsSumY* will rarely be exactly equal to zero due to the presence of noisy source signals that generate residue information to be coded. Nevertheless, as the *C4.5* algorithm tries to find a threshold for the attribute that minimises the overall classification error of the tree, a value different from zero can be selected for the attribute if such choice yields a smaller error. Figures B.8 and B.9 of Appendix B show that the thresholds chosen for the *AbsSumY* attribute in the decision trees corresponding to 16×16 and 32×32 CUs are 81 and 41, respectively, which are small numbers in comparison to their maximum values present in the training set (1149 and 3334 for 16×16 and 32×32 TUs, respectively).

Differently from *AbsSumY*, which presents one of the largest information gain levels among all attributes, *AbsSumU* and *AbsSumV* add little information to the decision trees. However, as using them does not incur in any extra computational complexity, since these are variables already computed and available during the encoding process, they are considered in the training of the decision trees.

Statistics for the *nonZeroCoeffY* attribute are shown in Fig. 6.14. The *nonZeroCoeffU* and *nonZeroCoeffV* attributes presented very similar statistics. Notice that the statistics for the cases when a TU has only zero-valued coefficients (i.e. *nonZeroCoeffY*=0) are identical to the statistics presented in Fig. 6.13 for the case when the sum of absolute luminance residues is equal to zero (i.e. *AbsSumY*=0). This happens because when the transform receives only zero values as input, it yields only zero-valued coefficients. Although *nonZeroCoeffY* is redundant with the *AbsSumY* attribute in this case, it still yields relevant information when its value is larger than zero. For example, Fig. 6.14a shows that when *nonZeroCoeffY* is larger than zero, 82.58 % of the 32×32 TUs are split into four 16×16 TUs and this information cannot be obtained with the *AbsSumY* attribute.

The *SingleCost* attribute is the R-D cost of encoding a TU as a whole block, i.e. without splitting it into four smaller TUs. Statistics for this attribute are presented in Fig. 6.15a for the specific case of 32×32 TUs and in Fig. 6.15b for 16×16 TUs. The

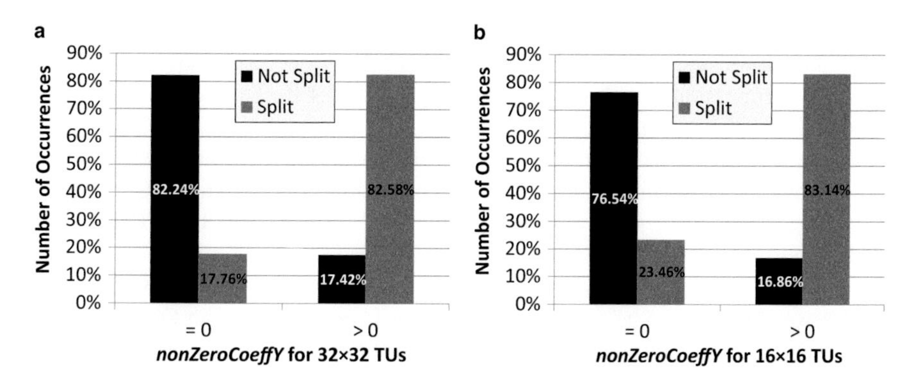

Fig. 6.14 Occurrence of (**a**) 32×32 and (**b**) 16×16 TUs split and not split into smaller TUs according to the number of nonzero luminance coefficients after the transform

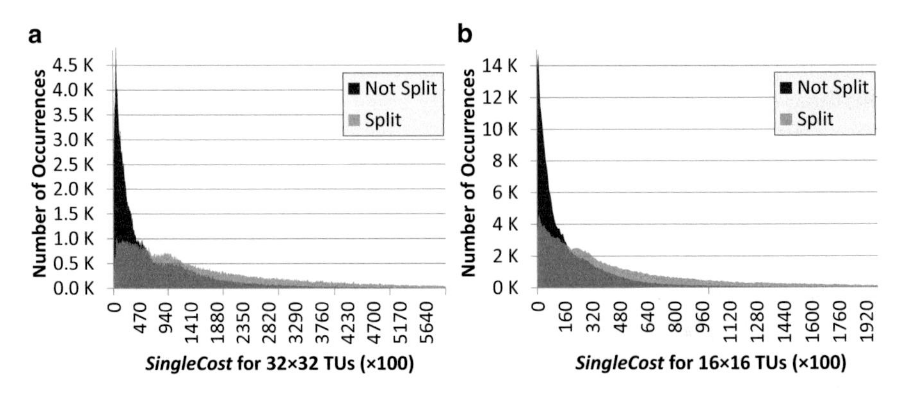

Fig. 6.15 Occurrence of (**a**) 32×32 and (**b**) 16×16 TUs split and not split into smaller TUs according to the R-D cost for the current TU depth

Table 6.6 Characteristics of trees trained for the RQT early termination

CU size	Depth	Test nodes	Leaves	Decision accuracy (%)	Incorrect depth (%)
32×32	10	15	16	83.4	8.9
16×16	8	17	18	80.7	12.1

figures show that there is a correlation between the TU splitting decision and the *SingleCost* value, since most of the non-split cases present very small R-D costs. Oppositely, for larger *SingleCost* values, the number of non-split TUs decreases and the split cases become the majority, although by a small amount.

The decision trees for 32×32 and 16×16 TUs are detailed in Table 6.6, which shows that their decision accuracies are 83.4 % and 80.7 %, respectively. As explained before in relation to the previous schemes, these decision accuracy results consider that splitting a TU when there is no need to do so is an error. However, the R-D efficiency is not harmed in such cases, since the best splitting option will be chosen through the exhaustive RDO evaluation procedure. Incorrect TU depth decisions occur only in the cases when the TU should be split into smaller TUs, but the process is early terminated after the whole TU is evaluated. The sixth column of Table 6.6 shows statistics regarding the incorrect depth decisions caused by incorrect early termination, which varies between 8.9 and 12.1 %. The two decision trees are detailed in the Appendix B of this book.

Figure 6.16 shows the 100th frame of the *BasketballDrill* video sequence, which was not used in the training of the decision trees, and its corresponding TUs for each CU. Large, medium and small circles represent 32×32, 16×16 and 8×8 TUs, respectively. Grey, blue and red circles represent the luminance, blue chrominance and red chrominance TUs, respectively. In areas where the TUs are not shown, no prediction residue was generated to be encoded. The sequence was encoded with QP 32 and the *Random Access* temporal configuration. The frame shown in Fig. 6.16a was encoded by the original HM encoder, while the frame in Fig. 6.16b was encoded by the HM encoder with the RQT early termination implemented. It is possible to perceive that there are few differences between the circles shown in Fig. 6.16a, b, which confirms that the early termination is capable of performing a correct RQT determination in most cases.

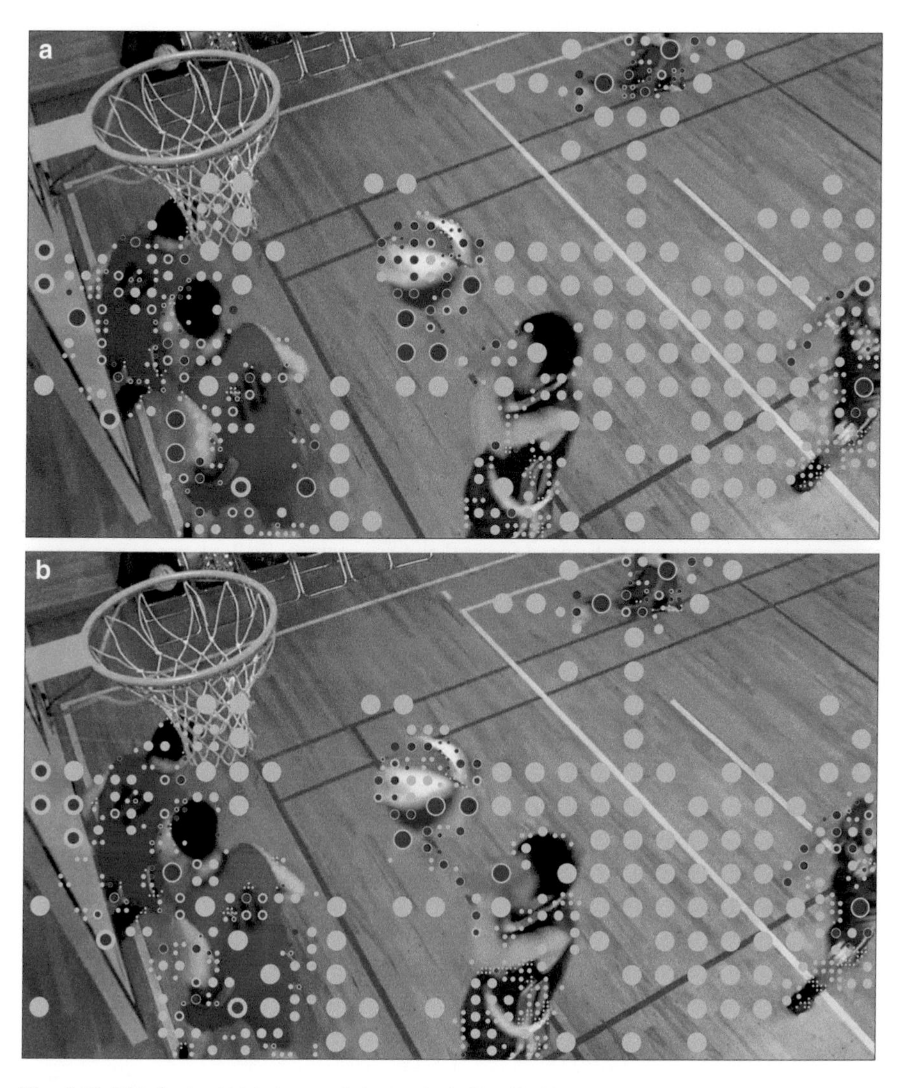

Fig. 6.16 TUs in the 100th frame of the *BasketballDrill* video sequence encoded with QP 32 by (**a**) the original HM encoder and the (**b**) HM encoder with the RQT early termination

6.6 Experimental Results

6.6.1 Experimental Setup and Methodology

In order to evaluate the performance of the three early termination methods proposed in this chapter, the respective decision trees were implemented in the HM encoder so as to build the seven schemes listed in Sect. 6.2. The HM encoder—version 12 (HM12)—was compiled with the *Microsoft Visual Studio C++ Compiler* and run on a clustered computer based on *Intel® Xeon® E5520* (2.27 GHz) processors running the *Windows Server 2008 HPC* operating system. Ten video sequences (*BasketballPass*, *BQSquare*, *BasketballDrill*, *ChinaSpeed*, *Kimono*, *SlideEditing*, *BQTerrace*, *Cactus*, *PeopleOnStreet* and *SteamLocomotive*), which are described in detail in Appendix A, were used to validate the early termination schemes and measure their R-D-C efficiency. As shown in Appendix A, these sequences differ from one another in terms of frame rate, bit depth, and spatial resolution. It is important to highlight that, in order to properly validate the early terminations, none of these test sequences was used in the training of the decision trees.

Compression efficiency was measured in terms of BD-rate and BD-PSNR, and the encoding times were obtained with the *Intel VTune Amplifier XE* software profiler [11]. BD-rate, BD-PSNR and encoding time variations were computed using the corresponding values obtained with the unmodified HM encoder as reference. All video sequences were encoded with QPs 22, 27, 32 and 37, using the *Random Access* temporal configuration.

It should be pointed out that the HM encoder is a software implementation of HEVC developed during the standardisation process for tests and documentation purposes, so that it is not optimised for real-time operation. Still, as the definition of the partitioning structures is a high-level decision which does not directly affect the operation performed at any particular HEVC encoder module (though it influences the number of encoding iterations performed in the RDO process), similar complexity reduction factors are expected to be achieved in other encoder implementations more efficient than HM. It is also important to notice that the current version of the HM encoder already includes several complexity reduction techniques, such as RMD for intra-frame prediction, the CBF-based early termination, an early CU termination algorithm and the early *SKIP* mode decision algorithm. The schemes presented in this chapter were implemented on top of all these methods, providing additional computational complexity reductions, and they were compared to the original HM encoder with all its built-in early terminations enabled.

The early terminations were first separately and then jointly implemented and evaluated. The next subsections present R-D-C results for the seven implementations listed in Sect. 6.2.

6.6.2 Rate-Distortion-Complexity Results for Single Schemes

Tables 6.7, 6.8 and 6.9 present R-D-C results for the three proposed schemes implemented separately. The results show that the PU early termination (Table 6.8) yields the largest reductions in computational complexity (*CCR* column), decreasing the total encoding time in a range from 37.4 to 68.1 % (on average, 49.6 %). These large reductions are explained by the fact that the scheme is applied to inter-predicted CUs, and as inter-frame prediction is the most time-consuming task in the HEVC encoder, the complexity reductions achieved with early terminated inter-predicted CUs have an important impact in the overall encoding complexity. The coding tree early termination scheme (Table 6.7) decreases the computational complexity in a

Table 6.7 R-D-C results for the coding tree early termination

Video sequence	BD-rate (%)	BD-PSNR (dB)	CCR (%)	Ratio BD-rate/CCR
BQSquare	–0.007	0.000	23.0	–0.030
BQTerrace	–0.008	0.000	28.4	–0.027
BasketballDrill	+0.425	–0.017	30.1	1.414
BasketballPass	+0.128	–0.006	23.9	0.534
Cactus	+0.207	–0.007	41.0	0.506
ChinaSpeed	+0.131	–0.007	29.2	0.449
Kimono	+0.552	–0.020	44.0	1.253
PeopleOnStreet	+0.089	–0.004	16.1	0.556
SlideEditing	+0.353	–0.057	71.3	0.495
SteamLocomotiveTrain	+0.971	–0.022	60.3	1.608
Average	**+0.284**	**–0.014**	**36.7**	**0.774**

Table 6.8 R-D-C results for the prediction unit early termination

Video sequence	BD-rate (%)	BD-PSNR (dB)	CCR (%)	Ratio BD-rate/CCR
BQSquare	+0.299	–0.014	45.7	0.655
BQTerrace	+0.091	–0.002	54.4	0.168
BasketballDrill	+0.491	–0.020	44.6	1.101
BasketballPass	+0.449	–0.020	42.5	1.055
Cactus	+0.401	–0.012	48.6	0.827
ChinaSpeed	+1.001	–0.051	47.1	2.127
Kimono	+0.689	–0.024	50.9	1.353
PeopleOnStreet	+1.021	–0.048	37.4	2.733
SlideEditing	+0.206	–0.030	68.1	0.302
SteamLocomotiveTrain	+0.969	–0.022	57.2	1.694
Average	**+0.562**	**–0.024**	**49.6**	**1.132**

Table 6.9 R-D-C results for the RQT early termination

Video sequence	BD-rate (%)	BD-PSNR (dB)	CCR (%)	Ratio BD-rate/CCR
BQSquare	+0.073	–0.003	8.9	0.820
BQTerrace	–0.086	+0.001	7.9	–1.093
BasketballDrill	+0.049	–0.002	6.5	0.753
BasketballPass	+0.033	–0.002	6.8	0.482
Cactus	+0.042	–0.001	7.2	0.575
ChinaSpeed	+0.131	–0.006	6.7	1.959
Kimono	+0.169	–0.006	5.6	2.988
PeopleOnStreet	+0.249	–0.012	6.2	4.022
SlideEditing	–0.244	+0.036	8.9	–2.739
SteamLocomotiveTrain	+0.132	–0.004	7.3	1.806
Average	**+0.055**	**0.000**	**7.2**	**0.758**

range from 16.1 to 71.3 % (on average, 36.7 %), while the RQT early termination (Table 6.9) yields a computational complexity reduction varying from 5.6 to 8.9 % (on average, 7.2 %).

It is possible to notice from the results shown in Table 6.7 that the CCR values provided by the coding tree early termination are correlated with the texture characteristics of the videos. Sequences with large homogeneous areas (e.g. *SlideEditing* and *SteamLocomotiveTrain*) are those which present the largest complexity reductions, because such areas are usually encoded with large CU sizes, allowing the early termination to halt the decision process in the first coding tree depths without significant loss of encoding performance and saving a significant amount of computation. High-resolution videos usually have larger numbers of homogeneous areas, but this is not always true. For example, *BQTerrace* and *PeopleOnStreet* are examples of high-resolution videos that do not present very large complexity reductions because they are composed of detailed texture. Videos with small spatial resolution (e.g. *BQSquare*, *BasketballDrill*, *BasketballPass*) also present smaller complexity reductions, since they are mostly encoded with small-sized CUs.

The largest complexity reductions achieved with the PU early termination (Table 6.8) are also observed in those videos with large homogeneous areas, since in these cases the decision process is halted more frequently after testing the largest PU size ($2N \times 2N$). However, different from the coding tree early termination, the complexity reductions achieved with the PU early termination also depend on the motion characteristics of the video. For example, the *BQTerrace* video sequence presents a larger complexity reduction with the PU early termination than with the coding tree early termination because it shows a scene in slow motion (mostly camera movement), which means that testing MSM or large PU modes in reference frames is usually enough to find a good prediction. On the other hand, fast-motion scenes with noncontinuous movement (e.g. *BasketballDrill*, *BasketballPass*) are those that present the smallest complexity reductions, since more modes need to be tested.

Notice that in some video sequences, a small BD-rate decrease was noticed instead of the expected increase. This is the case of the *BQSquare* and *BQTerrace* sequences in Table 6.7 and the *BQTerrace* and *SlideEditing* sequences in Table 6.9. A careful analysis of the encoding results for these sequences showed that in some areas of some frames, the early terminations led to the use of larger partitioning structures than those chosen by the original HM encoder. This is the case of the 11th frame in the *SlideEditing* sequence. The analysis also showed that the motion fields for this frame and following ones, obtained when using the early termination procedures, were more coherent than those obtained when using the original encoder. Even though these early decisions do not provide the best R-D efficiency possible for the current CTU, it appears that the bits saved through the increased use of large partitions, especially with MSM and *SKIP* modes, result in lower BD-rates. As these particular video sequences present homogeneous and moderate-motion characteristics, small or no image quality decreases are noticed due to these decisions, which, associated to the bit rate decrease, leads to the negative BD-rate values shown in the tables.

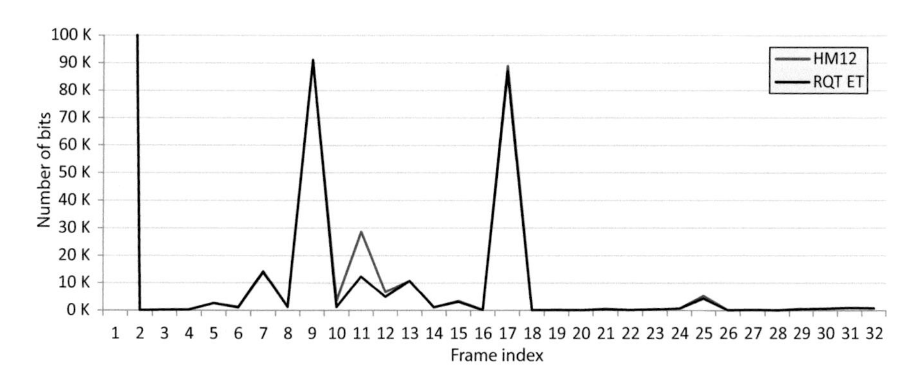

Fig. 6.17 Number of bits per frame in the *SlideEditing* sequence (QP 32) encoded with the original HM and the modified HM with the RQT early termination

To illustrate this explanation, the chart in Fig. 6.17 shows the number of bits per frame obtained when encoding the *SlideEditing* video (QP 32) with the original and the modified (*RQT ET*) HM encoders. The data shows that no differences are noticed in the number of bits between the two encoder versions, except for frame 11, where the modified encoder shows a much smaller number of bits than the original one. When analysing the whole video sequence, it was noticed that frame 11 is the one with the largest motion activity. A detailed analysis of this frame was then performed in order to understand the effect and its causes.

Figure 6.18 shows the fragment of the 11th frame in which most of the motion activity occurs. In Fig. 6.18a, the PU borders and MVs are shown for the fragment encoded with the original HM, while in Fig. 6.18b such information is omitted. Similarly, Fig. 6.18c, d shows the same area encoded with the modified HM. Blue and violet lines represent MVs referring to frames in *List 0* and *List 1*, respectively. By comparing Fig. 6.18a, c, it is possible to notice that the number of MVs and PUs in the fragment encoded with the original HM is much larger than in the modified version. In the specific case of this video, and especially in this region where most of the motion activity occurs, the early termination led to the choice of larger partitioning structures and MVs than in the original encoder, which finally resulted in a much smaller bit rate. As this is a very homogeneous video (computer-generated imagery), such larger partitioning structures and MVs led to the choice of more coherent motion fields than in the original encoder, which can be noticed in Fig. 6.18b, d by the quality of the encoded images. The circled area in Fig. 6.18b, for example, shows that the original encoder predicted the image area corresponding to the word "minor" by dividing it into several small PUs (see Fig. 6.18a), which were predicted from different areas of reference frames, resulting in something that reads as "mind". Oppositely, the same circled area in Fig. 6.18d shows that the modified encoder predicted the area as a single large PU with an MV pointing to a vertically offset position in a previous frame (*List 0*), resulting in the correct word "minor".

When compared to the six alternative partitioning structure configurations tested in Sect. 4.2, the three schemes proposed here achieve much better R-D-C efficiency.

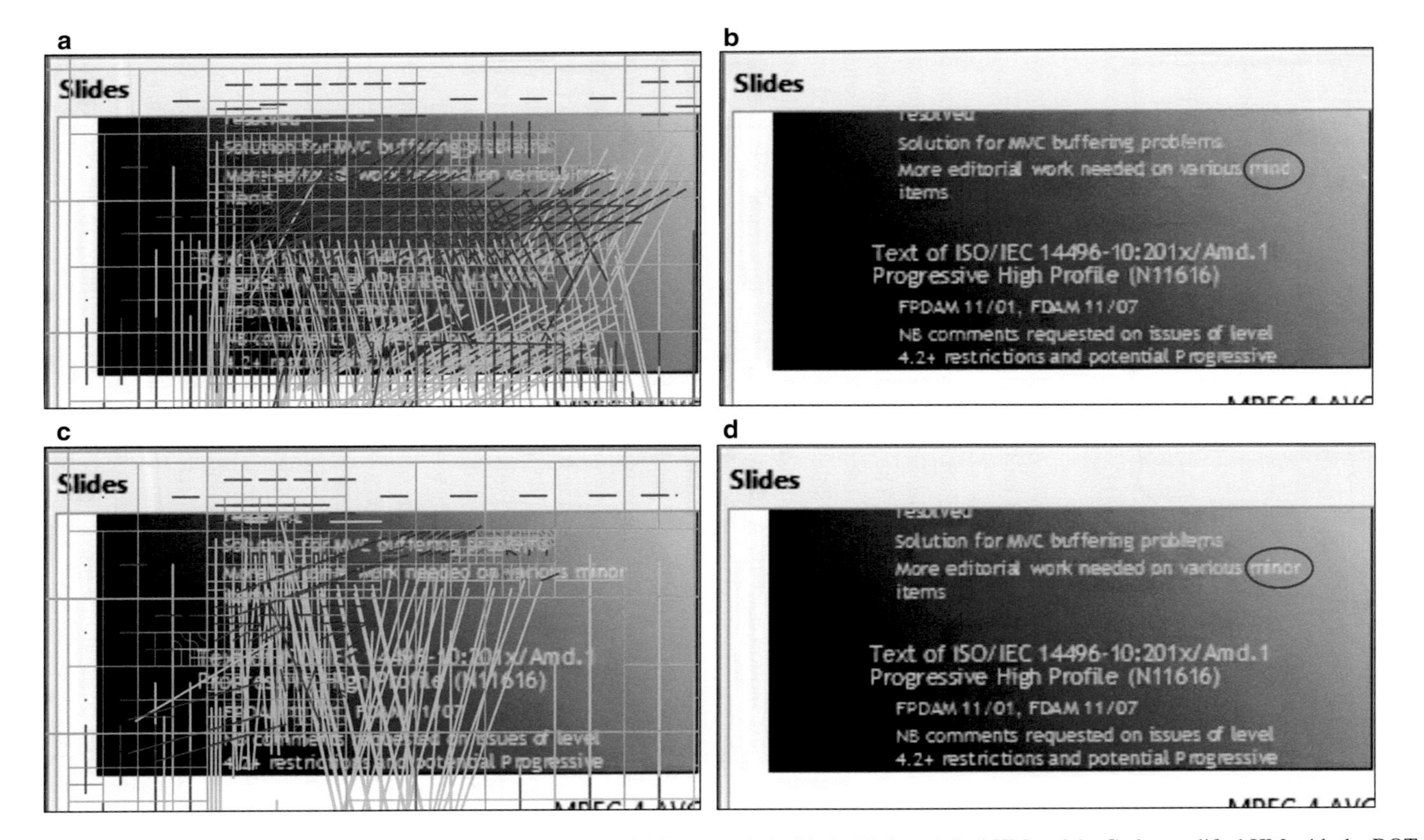

Fig. 6.18 Fragment of the 11th frame in the *SlideEditing* sequence (QP 32) encoded with (**a**, **b**) the original HM and (**c**, **d**) the modified HM with the RQT early termination

The alternative partitioning structure configuration of Sect. 4.2 that provided a complexity reduction closer to an early termination scheme proposed in this chapter, namely the PU early termination, is *PAR 3*. While that configuration provides a complexity reduction of 53.5 % in comparison to the original HM encoder (see Table 4.6), the PU early termination achieves a slightly smaller complexity reduction of 49.6 % (see Table 6.8). However, the BD-rate increase observed with *PAR 3* is 18.1 % (see Table 4.5), while the increase for the PU early termination is only 0.56 % (see Table 6.8). Clearly, the R-D-C efficiency of the method proposed in this chapter is much better. In fact, while the BD-rate/CCR ratio for those configurations in Sect. 4.2 varied between 3.2 and 44.2 (see Table 4.7), the three early terminations proposed in this chapter resulted in average ratios between 0.76 and 1.13, when implemented separately.

Figure 6.19 shows the R-D efficiency of the three proposed schemes for four video sequences encoded with four different QPs (22, 27, 32 and 37). In all charts of Fig. 6.19, the curves represent R-D results for the original encoder (*HM12*), the coding tree early termination (*CT ET*), the PU early termination (*PU ET*) and the RQT early termination (*RQT ET*) implementations, respectively. The charts of Fig. 6.19a, b show results for the *SlideEditing* and *BQTerrace* video sequences,

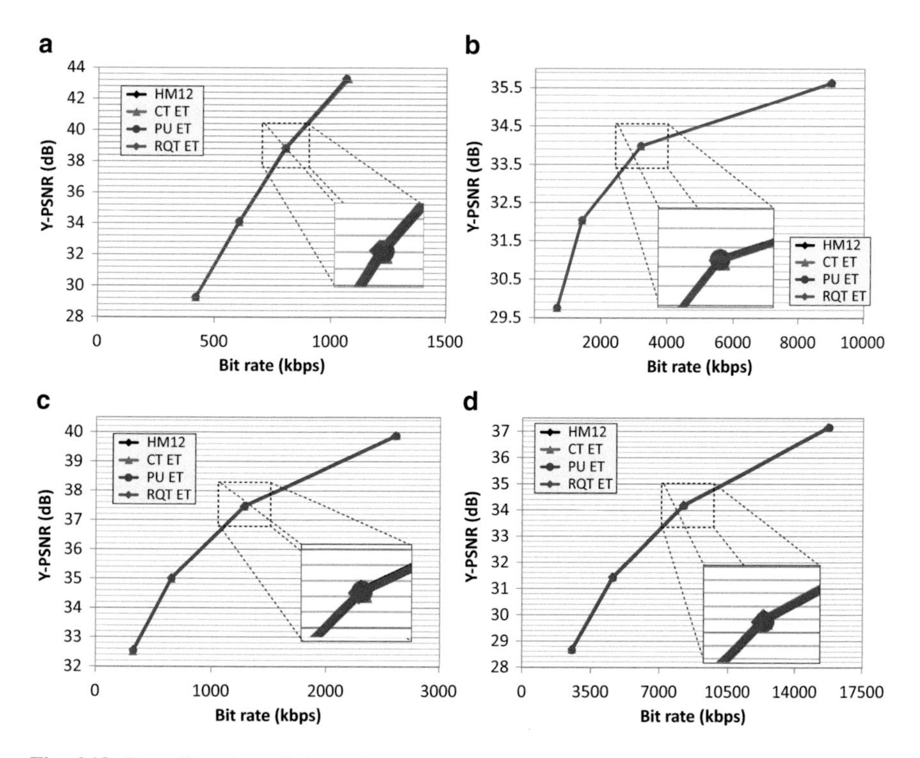

Fig. 6.19 Rate-distortion efficiency for the (**a**) *SlideEditing*, (**b**) *BQTerrace*, (**c**) *Kimono* and (**d**) *PeopleOnStreet* video sequences encoded with the original HM and the three early terminations implemented separately

which are those that presented the best R-D-C efficiency considering the three early terminations. The two remaining charts (Fig. 6.19c, d) are for the *Kimono* and *PeopleOnStreet* sequences, which are those that presented the worst R-D-C efficiency results. It is perceptible that the R-D efficiency achieved with the three proposed early terminations is very close to that of the original encoder, since the curves overlap in all charts presented in Fig. 6.19. A closer detailed look (300 % zoom boxes) shows that the curves overlap even in the worst-case video sequences.

6.6.3 Rate-Distortion-Complexity Results for Joint Schemes

The early terminations were jointly implemented in HM to investigate if by combining the different methods the individual complexity reductions added or if the use of some methods hindered the operation of the others, resulting in a total complexity reduction smaller than the sum of the partial reductions [4].

Tables 6.10, 6.11 and 6.12 present results for three schemes that use two early terminations in conjunction: coding tree and PU early terminations, PU and RQT early terminations and coding tree and RQT early terminations, respectively. Finally, results for an implementation that aggregates the three early terminations are presented in Table 6.13. As expected, the largest computational complexity reductions are achieved when all the early termination schemes are jointly implemented in HM. In this case, the computational complexity reductions vary from 46.4 % up to 87.6 % (65.3 %, on average, as shown in Table 6.13). Notice that the reduction achieved is not equivalent to the sum of the reductions achieved with each scheme separately. This happens due to the nature of the partitioning structures used in HEVC. For example, when constraining the number of R-D evaluations performed to decide the best coding tree configuration, the overall number of PU partitioning evaluations is also decreased, since PUs are defined within CUs. Similarly, the num-

Table 6.10 R-D-C results for joint coding tree and PU early terminations

Video sequence	BD-rate (%)	BD-PSNR (dB)	CCR (%)	Ratio BD-rate/CCR
BQSquare	+0.378	−0.017	54.3	0.697
BQTerrace	+0.283	−0.006	72.1	0.393
BasketballDrill	+1.295	−0.052	55.7	2.324
BasketballPass	+0.873	−0.038	50.9	1.717
Cactus	+0.999	−0.031	64.0	1.561
ChinaSpeed	+1.410	−0.072	56.3	2.505
Kimono	+2.745	−0.096	67.7	4.052
PeopleOnStreet	+1.463	−0.068	43.3	3.382
SlideEditing	+1.293	−0.190	86.6	1.493
SteamLocomotiveTrain	+2.573	−0.059	77.8	3.305
Average	**+1.331**	**−0.063**	**62.9**	**2.118**

Table 6.11 R-D-C results for joint PU and RQT early terminations

Video sequence	BD-rate (%)	BD-PSNR (dB)	CCR (%)	Ratio BD-rate/CCR
BQSquare	+0.378	–0.018	51.3	0.737
BQTerrace	+0.188	–0.005	58.9	0.320
BasketballDrill	+0.446	–0.018	48.2	0.925
BasketballPass	+0.735	–0.032	46.6	1.580
Cactus	+0.442	–0.013	52.7	0.839
ChinaSpeed	+1.206	–0.061	50.5	2.388
Kimono	+0.876	–0.031	53.9	1.627
PeopleOnStreet	+1.269	–0.059	40.8	3.113
SlideEditing	+0.075	–0.011	72.1	0.104
SteamLocomotiveTrain	+1.408	–0.032	60.9	2.311
Average	**+0.702**	**–0.028**	**53.6**	**1.311**

Table 6.12 R-D-C results for joint coding tree and RQT early terminations

Video sequence	BD-rate (%)	BD-PSNR (dB)	CCR (%)	Ratio BD-rate/CCR
BQSquare	+0.091	–0.004	30.3	0.301
BQTerrace	–0.028	0.000	56.4	–0.050
BasketballDrill	+0.453	–0.018	33.6	1.349
BasketballPass	+0.074	–0.003	29.0	0.255
Cactus	+0.284	–0.009	44.5	0.637
ChinaSpeed	+0.242	–0.012	33.1	0.733
Kimono	+0.839	–0.030	46.7	1.796
PeopleOnStreet	+0.396	–0.018	20.9	1.895
SlideEditing	+0.544	–0.072	73.2	0.744
SteamLocomotiveTrain	+0.553	–0.015	62.6	0.884
Average	**+0.345**	**–0.018**	**43.0**	**0.802**

Table 6.13 R-D-C results for joint coding tree, PU and RQT early terminations

Video sequence	BD-rate (%)	BD-PSNR (dB)	CCR (%)	Ratio BD-rate/CCR
BQSquare	+0.406	–0.019	59.0	0.689
BQTerrace	+0.282	–0.007	74.4	0.379
BasketballDrill	+1.182	–0.048	58.3	2.029
BasketballPass	+0.958	–0.042	54.1	1.773
Cactus	+1.006	–0.031	66.3	1.517
ChinaSpeed	+1.547	–0.078	58.8	2.629
Kimono	+3.013	–0.105	69.2	4.355
PeopleOnStreet	+1.740	–0.081	46.4	3.752
SlideEditing	+0.920	–0.134	87.6	1.050
SteamLocomotiveTrain	+2.493	–0.058	79.1	3.154
Average	**+1.355**	**–0.060**	**65.3**	**2.075**

ber of RQT evaluations also decreases because CUs are used as roots for RQTs, and so if there are less CUs with residues to be coded, there will be less RQTs whose optimal partition structure needs to be found.

In terms of R-D efficiency, the scheme with all early terminations presented the largest losses in comparison to the remaining versions, as expected. However, the average BD-rate increase of 1.36 % is still negligible in face of the considerable computational complexity reduction achieved, 65.3 %. For this reason, this configuration would be the most advisable solution for an implementation that admits a small loss in R-D efficiency. When compared to the alternative partitioning structure configurations tested in Sect. 4.2, the scheme presents much better R-D-C efficiency. *PAR 3* and *PAR 4* of Sect. 4.2 are the configurations that provided a complexity reduction closer to the achieved with the scheme that implements all early terminations. The two configurations provide a complexity reduction of 53.5 % and 74.1 %, respectively (see Table 4.6), while the joint early terminations achieve a reduction of 65.3 % (see Table 6.13). Nevertheless, the BD–rate increases of *PAR 3* and *PAR 4* are 18.1 and 36 % (see Table 4.5), respectively, while the increase for the joint early termination scheme is only 1.36 % (see Table 6.13).

Figure 6.20 shows R-D efficiency results for the four joint implementations described above. In all charts of Fig. 6.20, the curve labelled as *HM12* represents

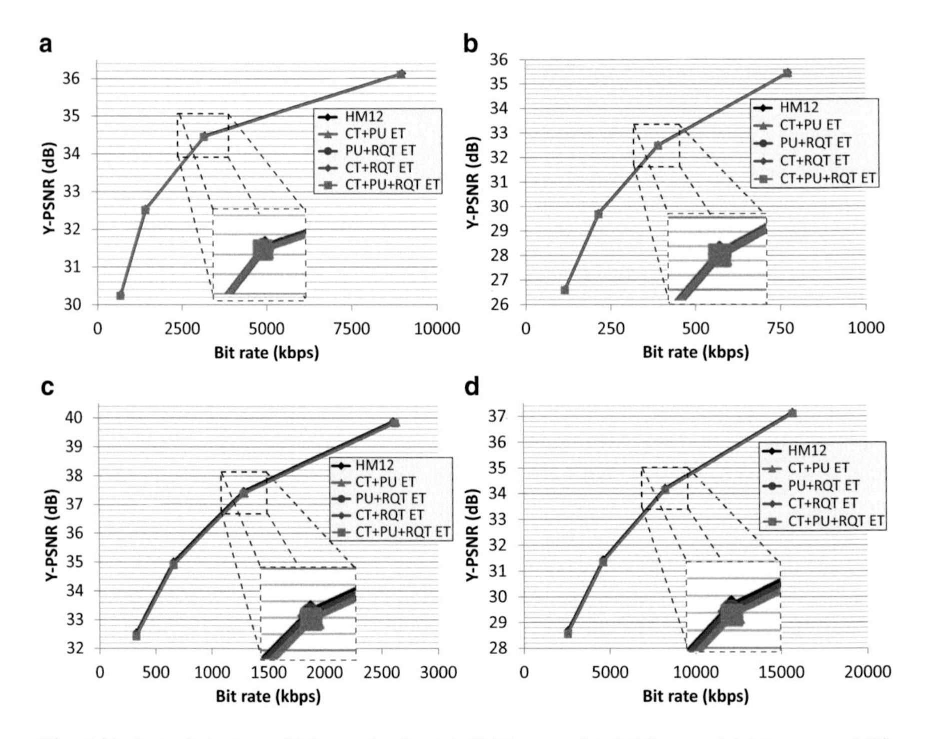

Fig. 6.20 Rate-distortion efficiency for the (**a**) *BQTerrace*, (**b**) *BQSquare*, (**c**) *Kimono* and (d) *PeopleOnStreet* video sequences encoded with the original HM and the four joint early termination schemes implemented

R-D results for the original encoder, while the curves labelled as *CT+PU ET*, *PU+RQT ET*, *CT+RQT ET* and *CT+PU+RQT ET* represent R-D results for the joint coding tree and PU early terminations, the joint PU and RQT early terminations, the joint coding tree and RQT early terminations and the joint implementation with the three early terminations, respectively. The charts in Fig. 6.20a, b show results for the *BQTerrace* and *BQsquare* sequences, respectively, which presented the best R-D-C results in the four joint schemes. Figure 6.20c, d shows results for the *Kimono* and *PeopleOnStreet* sequences, which presented the worst R-D-C efficiency results. As in the previous comparisons for the schemes implemented separately, the R-D efficiency achieved in the joint cases is also very close to that of the original encoder. Even in the worst-case videos, a detailed look (300 % zoom) shows that the curves are very close.

6.7 Results and Discussion

To quantify the quality of the methods proposed in this section, we compared their performance with those of the best performing HEVC complexity reduction methods reviewed in Chap. 3, which also operate through modifications in the partitioning structure decision process. Only those works that provide comparable results (i.e. BD-rate and complexity reduction values using the original HM encoder as reference) were considered in the analysis. All the compared works were also tested with the *Random Access* configuration, QPs 22, 27, 32 and 37, and were tested for at least seven video sequences with at least four different spatial resolutions, except for [12], which was kept in the comparisons because it was the only comparable work in its category.

Table 6.14 presents the results of these experiments in terms of BD-rate, computational complexity reduction (*CCR* column) and the ratio between the two values for all compared works. Since each method under evaluation presents different values for BD-rate and complexity reduction, the ratio between these two quantities (BD–rate/CCR) was used to permit comparisons of the competing complexity reduction methods in terms of R-D efficiency loss per computational complexity saved. Using this performance indicator, a method which presents a given BD–rate and CCR is ranked higher than some other methods with higher BD-rate but equal (or smaller) CCR (i.e. higher BD-rate/CCR ratio). The methods are grouped according to the partitioning structure that is constrained to achieve the desired complexity reduction. To ease the comparisons, the average results for the schemes proposed in this chapter are also reproduced at the end of each group. We can conclude from the data in Table 6.14 that all the schemes proposed in this chapter achieve better BD-rate/CCR ratios than the competing methods in their corresponding category. Even though an extensive research has been done in the available literature, no works have been found for comparisons in the categories *PU+RQT ET*, *CT+RQT ET* and *CT+PU+RQT ET*.

Table 6.14 Comparisons with related works

Category	Reference	BD-rate (%)	CCR (%)	Ratio BD-rate/CCR
CT ET	Seunghyun [13]	+0.6	50	1.20
	J.-Hyeok [14]	+1.2	48	2.50
	Goswami [15]	+1.7	38	4.42
	Jian [16]	+1.9	43	4.42
	Proposed	**+0.28**	**37**	**0.77**
PU ET	Vanne [17]	+1.3	51	2.55
	Zhao [18]	+5.9	50	11.8
	Khan [19]	+1.3	44	2.88
	Proposed	**+0.56**	**50**	**1.13**
RQT ET	Yunyu [12]	+1.4	22	6.36
	Proposed	**+0.05**	**7.2**	**0.76**
CT+PU ET	Wei-Jhe [20]	+5.1	43	5.11
	Liquan [21]	+1.5	42	3.55
	Xiaolin [22]	+1.9	41	4.54
	Xiaolin [23]	+1.4	45	3.11
	Proposed	**+1.33**	**63**	**2.12**
PU+RQT ET	**Proposed**	**+0.7**	**54**	**1.31**
CT+RQT ET	**Proposed**	**+0.34**	**43**	**0.80**
CT+PU+RQT ET	**Proposed**	**+1.36**	**65**	**2.07**

In the coding tree early termination category, the best related work [13] has a BD-rate/CCR ratio equal to 1.2, which is still larger than the ratio of the scheme proposed in Sect. 6.3 (0.77). Besides, the method in [13] is only applicable to intra-predicted CUs, which means that its complexity reductions are only achieved in *All Intra* temporal configurations. In the PU early termination category, all related works present a ratio at least twice larger than that obtained with the PU early termination scheme proposed in this chapter. Moreover, the PU early termination scheme presented in this book achieves a computational complexity reduction equal to or greater than that obtained in the related works. Finally, although the only comparable RQT early termination work achieves complexity reduction levels larger than those of the proposed RQT early termination, the cost of such reductions in terms of R-D efficiency loss is too large, producing a BD–rate/CCR ratio equal to 6.36, which is 8.4 times larger than the ratio of scheme proposed in Sect. 6.5 (0.76).

The four joint early termination schemes proposed in this chapter are listed in the last lines of Table 6.14 and compared to joint coding tree and PU early termination methods found in the literature. We can observe that all joint schemes proposed in this chapter achieve better results in both terms of R–D efficiency and computational complexity reduction. The proposed joint scheme that integrates the three early terminations achieves complexity reductions that largely exceed those obtained in other related works, while still maintaining a low BD-rate increase and, consequently, smaller (i.e. better) BD-rate/CCR ratios than those achieved by the comparable joint implementations.

6.8 Conclusions

This chapter presented a set of early termination schemes that reduce the computational complexity of the HEVC encoding process. All the schemes were developed making use of DM tools for the construction of decision trees that exploit intermediate encoding results to decrease the computational complexity involved in the decision of the best coding tree, PU and RQT structures. One single early termination scheme was separately developed and implemented in the HM software for each partitioning structure decision after performing an extensive analysis on the information gain yield by each possible attribute. The three implementations were then combined in pairs and finally all together in joint schemes, aiming at further reducing the HEVC computational complexity.

The effectiveness of the approach and the performance of the decision trees were validated through extensive experiments using a set of video sequences different from those used in the training phase. Experimental results have shown that an average complexity reduction of 65 % can be achieved when the three early termination schemes are jointly implemented, with a compression efficiency loss of only 1.36 % (BD-rate). Complexity reductions that go beyond those achieved in any other previously published comparable works were achieved through the proposed early termination methods, with smaller losses in terms of compression efficiency.

The proposed schemes do not require any computationally intensive operations to be added to the HEVC encoding process, and the decision trees use only intermediate encoding results computed during normal HEVC encoding. Therefore, these early termination schemes can be seamlessly incorporated to any other HEVC encoder implementation with very small increase in the computational burden of the encoding. For reference, all trained decision trees were made available in Appendix B as supplemental material to this chapter.

References

1. G. Correa, P. Assuncao, L. Agostini, L.A. da Silva Cruz, A Method for Early-Splitting of HEVC Inter Blocks Based on Decision Trees, in *European Signal Processing Conference (EUSIPCO 2014)*, Lisbon, Portugal, 2014
2. G. Correa, P. Assuncao, L. Agostini, L.A. da Silva Cruz, Four-step Algorithm for Early Termination in HEVC Inter-frame Prediction based on Decision Trees, in *Visual Communications and Image Processing (VCIP 2014)*, Valleta, Malta, 2014
3. G. Correa, P. Assuncao, L. Agostini, L.A. da Silva Cruz, Classification-Based Early Termination for Coding Tree Structure Decision in HEVC, in *IEEE International Conference on Electronics, Circuits, and Systems (ICECS 2014)*, Marseille, France, 2014.
4. G. Correa, P. Assuncao, L. Agostini, L.A. da Silva Cruz, Fast HEVC encoding decisions using data mining. IEEE Trans Circuits Syst Video Technol **24**, 660–673 (2015)
5. J.R. Quinlan, *C4.5: Programs for Machine Learning* (Morgan Kaufmann Publishers, Burlington, 1993)
6. M. Hall, E. Frank, G. Holmes, B. Pfahringer, P. Reutemann, I.H. Witten, The WEKA data mining software: an update. SIGKDD Explor Newsl **11**, 10–18 (2009)

7. ISO/IEC-JCT1/SC29/WG11, Common test conditions and software reference configurations, Geneva, Switzerland, 2012
8. N. Japkowicz, The Class Imbalance Problem: Significance and Strategies, in *Proceedings of the 2000 International Conference on Artificial Intelligence (IC-{AI}'2000)*, 2000, pp. 111–117
9. T.M. Cover, J.A. Thomas, *Elements of Information Theory (Wiley Series in Telecommunications and Signal Processing)* (Wiley-Interscience, Hoboken, NJ, 2006)
10. S. Kullback, R.A. Leibler, On Information and Sufficiency, pp. 79–86, 1951/03 1951
11. *VTune™ Amplifier XE from Intel.* Available: http://software.intel.com/en-us/articles/intel-vtune-amplifier-xe/
12. Y. Shi, Z. Gao, X. Zhang, Early TU Split Termination in HEVC Based on Quasi-Zero-Block, in *3rd International Conference on Electric and Electronics*, 2013
13. C. Seunghyun, K. Munchurl, Fast CU splitting and pruning for suboptimal CU partitioning in HEVC intra coding. Circuits Syst Video Technol IEEE Trans **23**, 1555–1564 (2013)
14. L. Jong-Hyeok, P. Chan-Seob, K. Byung-Gyu, Fast coding algorithm based on adaptive coding depth range selection for HEVC, in *Consumer Electronics—Berlin (ICCE-Berlin), 2012 IEEE International Conference on*, 2012, pp. 31–33
15. Kalyan Goswami, Byung-Gyu Kim, Dong-San Jun, Soon-Heung Jung, J. S. Choi, Early Coding Unit (CU) Splitting Termination Algorithm for High Efficiency Video Coding (HEVC), *to be published in ETRI Journal,* 2014
16. X. Jian, L. Hongliang, W. Qingbo, M. Fanman, A fast HEVC inter CU selection method based on pyramid motion divergence. Multimedia IEEE Trans **16**, 559–564 (2014)
17. J. Vanne, M. Viitanen, T. Hamalainen, Efficient mode decision schemes for HEVC inter prediction. Circuits Syst Video Technol IEEE Trans **24**, 1579–1593 (2014)
18. T. Zhao, Z. Wang, S. Kwong, Flexible mode selection and complexity allocation in high efficiency video coding. Selected Topics Signal Process IEEE J **7**, 1135–1144 (2013)
19. M.U.K. Khan, M. Shafique, J. Henkel, An Adaptive Complexity Reduction Scheme with Fast Prediction Unit Decision for HEVC Intra Encoding, in *IEEE International Conference on Image Processing (ICIP)*, 2013
20. H. Wei-Jhe, H. Hsueh-Ming, Fast coding unit decision algorithm for HEVC, in *Signal and Information Processing Association Annual Summit and Conference (APSIPA), 2013 Asia-Pacific*, 2013, pp. 1–5
21. S. Liquan, L. Zhi, Z. Xinpeng, Z. Wenqiang, Z. Zhaoyang, An effective CU size decision method for HEVC encoders. Multimedia IEEE Trans **15**, 465–470 (2013)
22. S. Xiaolin, Y. Lu, C. Jie, Fast coding unit size selection for HEVC based on Bayesian decision rule, in *2012 Picture Coding Symposium*, 2012, pp. 453–456
23. X. Shen, L. Yu, CU splitting early termination based on weighted SVM. EURASIP J Image Video Process **2013**, 4 (2013)

Chapter 7
Complexity Reduction and Scaling Applied to HEVC Encoding Time Control

The main goal of this chapter is to combine the findings and methods presented in Chaps. 4, 5 and 6 to build a control system that dynamically adjusts the encoder operation to keep the encoding time under a specific target. This system will be used as a showcase for the applicability of the major contributions described in the previous chapters of this book.

Chapter 4 presented an R-D-C analysis that identified a set of tools and parameter settings that have a large influence in the overall encoding computational complexity and showed that large complexity reductions can be achieved with small compression efficiency loss by choosing judiciously those settings. Chapter 4 has also shown that the HEVC frame partitioning structures are responsible for a very large share of the encoding complexity. These observations motivated the introduction of a set of complexity scaling and reduction methods described in Chaps. 5 and 6 that operate by constraining the decision process followed to arrive at the optimal topology of these structures in R-D sense.

However, there is a remaining important problem of practical nature related to the subject of this book: how can we combine the best low-complexity encoding tool configurations and the proposed complexity scaling and reduction algorithms to guarantee that encoding times are kept below a predefined target? This problem is addressed in this chapter, which presents an example of a system designed to control the encoding time per GOP based on the computational complexity reduction and scaling methods proposed in previous chapters.

The design of this encoder control system starts by the definition of a set of encoder operational configurations that combine different arrangements of encoding tools (based on the findings of Chap. 4) and different schemes for early terminating the partitioning structure decision process (based on the decision trees of Chap. 6). Then, each possible configuration is tested in an offline training process that is performed to derive the best configuration options in terms of R-D-C efficiency. These configurations are then sorted in descending order of encoding time. The encoding time control algorithm is then implemented using these configurations to adjust, at a medium-granularity level, the time spent by the HEVC encoder

G. Corrêa et al., *Complexity-Aware High Efficiency Video Coding*,
DOI 10.1007/978-3-319-25778-5_7

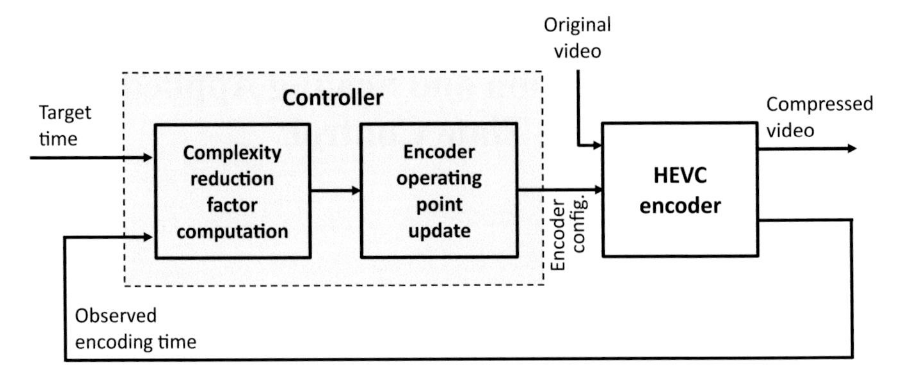

Fig. 7.1 Encoding time control system integrated with the HEVC encoder

to process each GOP. Once an operational configuration is chosen, a finer adjustment of the encoding time control is performed based on the CCUPU method presented in Chap. 5. This finer control is done by changing the number of constrained CTUs per frame, which allows achieving encoding times per GOP closer to the target time.

The high-level diagram of the encoding time control system proposed in this chapter is presented in Fig. 7.1. The inputs of the encoding time control are a target time per GOP, which might be specified from an external entity like an operation system process scheduler and the encoding time of previously encoded GOPs (or a function of previous GOPs encoding times). The necessary complexity reduction factor is calculated according to the difference or ratio of these two input values. Based on this factor, an encoder operating point is chosen in such a way that the encoding time is adjusted towards the desired value.

Section 7.1 explains how the encoder operating points were obtained and Sect. 7.2 presents the proposed encoding time controller. Experimental results that illustrate the control system operation and quantify its performance are presented and analysed in Sect. 7.3.

7.1 Finding the Encoder Operating Points

As previously explained, improved R-D efficiency was achieved through the addition of new tools and encoding features in HEVC, so that the resulting encoder supports several configurations at the cost of different levels of computational complexity, as shown in Chap. 4. If computational complexity is to be considered in the selection of encoding parameters, a new dimension must be added to the RDO problem, resulting in an R-D-C optimisation (RDCO). However, RDCO is a very difficult task to be performed in real-time encoding, mainly because differently from the rate (R) and distortion (D) variables, computational complexity cannot be retrieved once spent, so that it is not possible to go back and try another parameterisation without incurring in more computational cost. On the other hand, not testing all

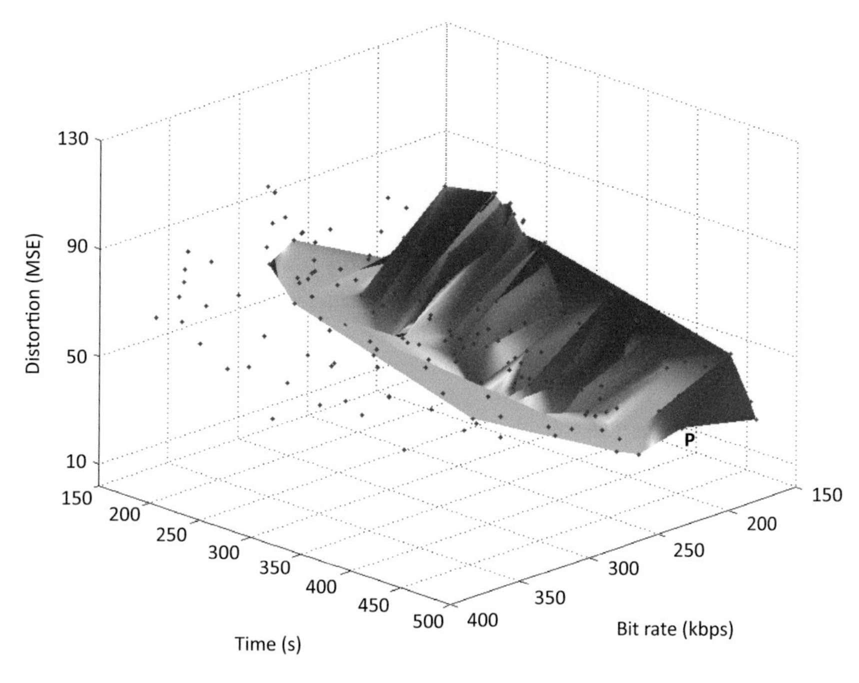

Fig. 7.2 Example of R-D-C space and 3-D Pareto frontier for a set of 240 encoding parameter configurations

possibilities can lead to a significant decrease in the compression efficiency if a careful selection of tested configurations is not performed.

Figure 7.2 shows the R-D-C space obtained when encoding a certain video sequence using a set of 240 different configurations. Each point in the chart represents the R, D and C values of a given configuration in terms of bit rate (kbps), distortion (MSE) and encoding time (seconds), respectively. If a point is not outscored by any other in the three terms (R, D and C), it is said to be a non-dominated point. For example, take the point labelled as *P* in Fig. 7.2. There is no other point in the chart that possesses, simultaneously, a smaller distortion, a smaller bit rate and a smaller encoding time than *P*, and for this reason, it is called a non-dominated point. All non-dominated points in Fig. 7.2 are shown as red points, and they correspond to the so-called Pareto frontier,[1] which means that they are those that present the best average R-D-C efficiency among all tested cases and thus should be selected for use in an R-D-C optimised encoding time control system. The next subsections explain how these best-performing points were obtained and utilised in the control system proposed in this chapter to adjust the encoding time per GOP.

[1] The Pareto frontier is composed of every point that is Pareto-efficient (i.e., not dominated by any other alternative). A point *A* is said to dominate *B* if *A* outscores *B* regardless of the trade-off between *A* and *B*. In other words, if *A* is better and cheaper than *B*, *A* dominates *B*. In the specific case of Fig. 7.2, a configuration *A* dominates another configuration *B* if it results in a smaller distortion, in a smaller bit rate and in a smaller encoding time than *B*. All configurations that are not dominated by any other are thus called Pareto-efficient configurations.

7.1.1 Parameters Selection for R-D-C Analysis

In order to implement the encoding time control system presented in this chapter, the encoding configurations that yield the best R-D efficiency for a given computational complexity were selected through an extensive set of offline evaluations. Even though the HM encoder presents a large number of configuration parameters, only those identified in Chap. 4 as having the largest computational complexity impact were selected for the experimental study described in this section. Besides those parameters, the three single early termination schemes proposed in Chap. 6 were also included in the R-D-C analysis, so that three parameters were added to the HM encoder representing the activation or deactivation state of each early termination scheme. The R-D-C analysis could include several other encoding parameters of HM, but, as shown later in this chapter, a limited subset of them had to be chosen with care to make this experimental analysis feasible. Expanding the set of parameter to larger numbers would call for a very large number of experimental evaluations of encoder performance due to the rapid growth of the number of configuration possibilities with the number of parameters.

Section 4.1.2 examined 17 configurations and identified seven which, when used, resulted in an increase of the encoding computational complexity in a range from 5 to 53 %, depending on the video sequence used (see Fig. 4.1). The remaining configurations did not alter the encoding computational complexity significantly. The seven most complex configurations in Sect. 4.1.2 are identified as *TEST 2–TEST 7* and *TEST 14*, and they correspond, respectively, to modifications in the *Inter 4×4*, *Search Range*, *Bi-prediction Refinement*, *Hadamard ME*, *Fast Encoding*, *Fast Merge Decision* and *AMP* encoding parameters. In Sect. 4.1.4, the accumulated effect of enabling each tool was also analysed, and it was concluded that the same seven parameters were responsible for the largest increases in computational complexity when compared to a baseline configuration (see Fig. 4.4).

Due to their large impact in computational complexity, these seven parameters are the best candidates for use in the construction (by combinations and variations of their values) of the configurations to be used in the R-D-C analysis presented in this section. However, only three of them could actually be analysed[2]: the *Search Range* (*SR*), the *Bi-prediction Refinement* (*BPR*) and the *Hadamard ME* (*HME*). They are respectively referred as *SR*, *BPR* and *HME* from now on in this chapter. Similarly, the three early termination schemes described in Chap. 6, namely, the *coding tree early termination* (*CTET*), the *prediction unit early termination* (*PUET*)

[2] The *Inter 4×4* parameter was not included in the R-D-C analysis because it is not supported in the most recent HM versions. The configurations modifying the *Fast Encoding* and the *Fast Merge Decision* parameters in Chap. 4 actually disabled them (thus increasing the computational complexity and decreasing R-D efficiency), so that they are always enabled in the CTC-based configurations used in the R-D-C analysis presented in this chapter. The *AMP* parameter controls the use of *asymmetric motion partitions*, which is already controlled by the *prediction unit early termination* (*PUET*).

Table 7.1 Encoding configurations tested in the R-D-C analysis

Configuration	SR	BPR	HME	CTET	PUET	RQTET
1	64	4	On	Off	Off	Off
2	32	4	On	Off	Off	Off
3	16	4	On	Off	Off	Off
4	8	4	On	Off	Off	Off
…	…	…	…	…	…	…
31	64	4	On	On	Off	Off
…	…	…	…	…	…	…
240	4	1	Off	On	On	On

and the *residual quadtree early termination* (*RQTET*), were used to define the configurations. In the experiments that will be described shortly, these six parameters can take values from the following sets:

- $SR \in \{64, 32, 16, 8, 4\}$
- $BPR \in \{4, 2, 1\}$
- $HME \in \{\text{on, off}\}$
- $CTET \in \{\text{on, off}\}$
- $PUET \in \{\text{on, off}\}$
- $RQTET \in \{\text{on, off}\}$

Table 7.1 shows part of the encoding configurations that were created by modifying the value of each parameter, one at a time. Every parameter value was tested with all possible combinations of values of the remaining parameters, totalising 240 encoding configurations, all of which are detailed in Appendix C.

7.1.2 Operating Points at the Pareto Frontier

Each one of the 240 configurations mentioned in the previous section was used to encode 150 frames of 10 high-resolution (1080p) video sequences with QPs 22, 27, 32, 37 and the *low delay* configuration, totalising 9600 encodings. The ten high-resolution sequences used in such analysis are the *BasketballDrive*, *BQTerrace*, *Cactus*, *Kimono1*, *ParkScene*, *Poznan_CarPark*, *Poznan_Hall1*, *Poznan_Street*, *Shark1*, *Tennis*, detailed in Appendix A of this book. Notice that even though HM was not designed for real-time operation, it is currently the most complete software implementation of the HEVC encoder, including all tools defined in the standard. For this reason, it was used to perform a thorough analysis of the R-D-C efficiency of the most time-consuming configurations. BD-rate and BD-PSNR values were calculated for each configuration using the most complex configuration as reference (i.e. configuration 1: $SR = 64$, $BPR = 4$, $HME = \text{on}$, $CTET = \text{off}$, $PUET = \text{off}$, $RQTET = \text{off}$). As encoding time and computational complexity are usually considered directly proportional, computational complexity was also measured in terms of encoding time and normalised with reference to configuration 1.

To model constant bit rate and constant video quality encoding modes, two different analyses were performed in two-dimensional (2D) spaces: one examining BD results in a rate-complexity (R-C) space and another examining BD results in a distortion-complexity (D-C) space. Figure 7.3a, b shows the R-C and the D-C spaces with the (average) BD and computational complexity results for the 240 configurations. Since configuration 1 was used as reference, it appears in the right-most corner of the plots, with the best encoding performance and the largest computational complexity. Detailed results in terms of bit rate, PSNR and encoding time for each configuration are presented in Appendix C.

The best encoding configurations were identified as those corresponding to the points that belong to the Pareto frontier (i.e. non-dominated points) in both the R-C and D-C spaces. In the specific case of Fig. 7.3a, a configuration A dominates another configuration B if it results in a smaller BD-rate and in a smaller computational complexity than B. Similarly, in Fig. 7.3b, a configuration A dominates a configuration B if it yields a larger BD-PSNR (i.e., smaller performance degradation) and a smaller computational complexity than B. This way, by selecting the points that belong to the Pareto frontier, it is possible to identify those configurations that yield the best average results in terms of R-D-C efficiency within the analysed scenario. In both Fig. 7.3a, b, configurations that belong to the Pareto frontier in the R-C space are represented with squares and configurations that belong to the Pareto frontier in the D-C space are represented with circles. The remaining configurations are represented with the × symbol.

Notice that several configurations belong simultaneously to the two Pareto frontiers (circles inside squares in the charts of Fig. 7.3). Such cases represent the best encoding options for a determined computational complexity both in terms of bit rate and image quality, considering all the options of the analysed set. Even when the configurations belong to only one Pareto frontier (either a square or a circle), they perform well in the other space. For example, all configurations that belong to the D-C space Pareto frontier (circles) are located close to the lower convex hull of the R-C space (see Fig. 7.3a). Similarly, most configurations that belong to the R-C space Pareto frontier (squares) are close to the upper convex hull of the D-C space (see Fig. 7.3b). This happens because the two BD measures used in the analysis take into account variations in both the bit rate and PSNR.

Figure 7.3 also shows that the encoding performance diminishes with the decrease of the relative computational complexity from 1 downwards 0. Initially, very small encoding performance decreases are noticed, especially when the relative computational complexity is above 0.82. From 0.82 down to 0.43, the encoding performance diminishes gradually until it reaches a BD-rate increase of 1.11 % and a BD-PSNR decrease around 0.026 dB. Configurations with relative complexity below 0.43 result in encoding performance losses considered too large for the small levels of complexity reduction achieved, so that they were not selected for the control system proposed in the next section.

Table 7.2 shows the 15 selected configurations that belong to the Pareto frontier of the plots shown in Fig. 7.3. The table shows the value for each parameter and the resulting normalised computational complexity. The next section will present the

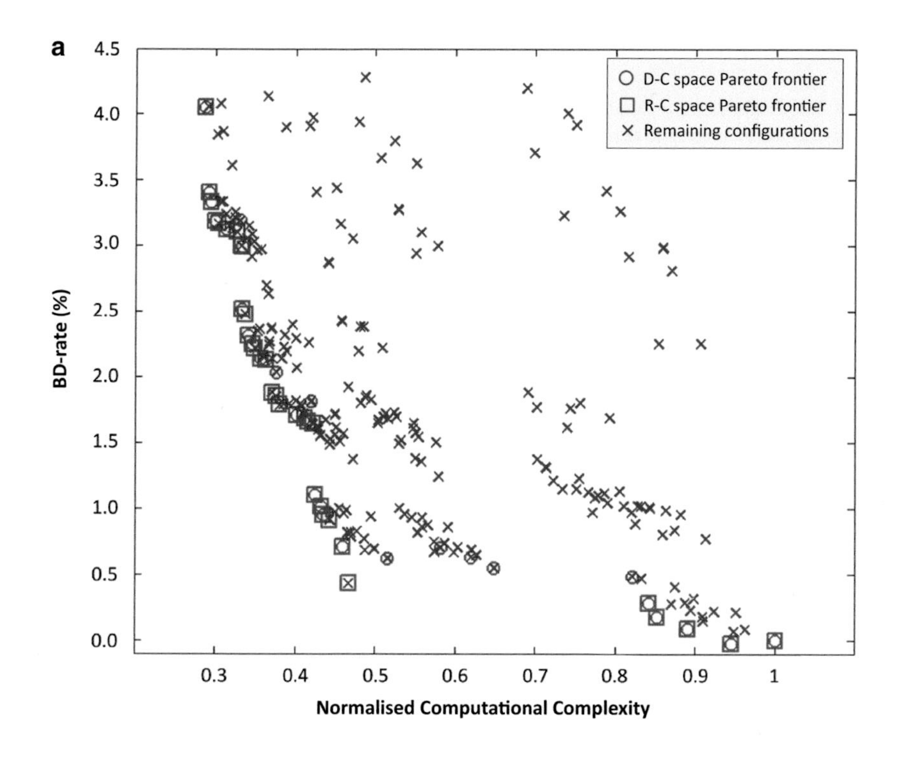

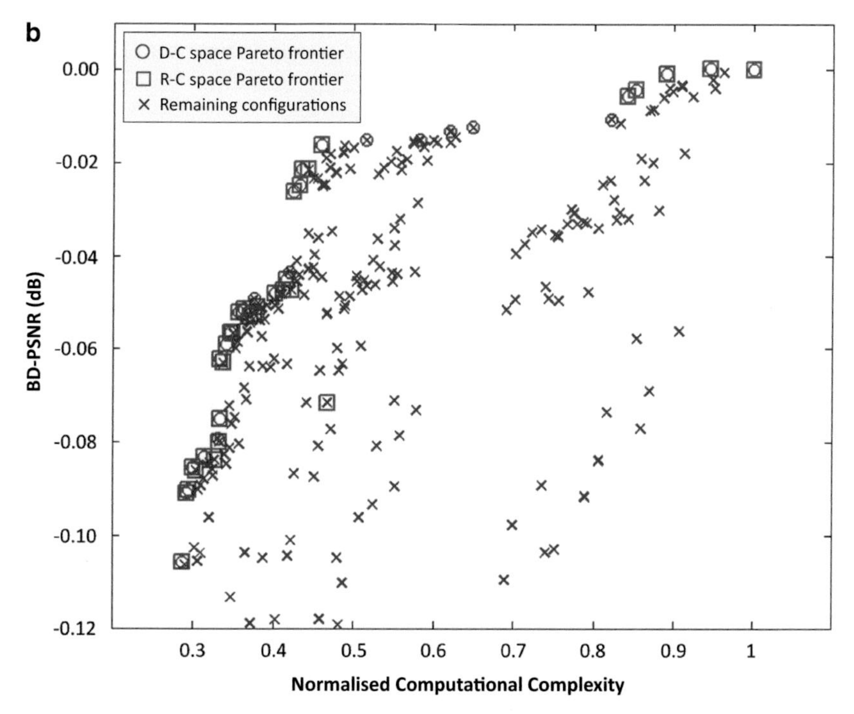

Fig. 7.3 (**a**) R-C and (**b**) D-C average results for the 240 configurations tested. *Squares* represent Pareto-efficient configurations in the R-C space, *circles* represent Pareto-efficient configurations in the D-C space and × *symbols* represent the remaining configurations

Table 7.2 Parameters, normalised computational complexity, BD-PSNR and BD-rate for the configurations belonging to the Pareto frontier

C	O	SR	BPR	HME	CT ET	PU ET	TU ET	Complexity	BD-PSNR	BD-rate
C_1	O_1	64	4	1	0	0	0	1	0	0
C_2	O_2	32	4	1	0	0	0	0.945	0	0.002
C_3	O_{122}	32	4	1	0	0	1	0.890	–0.001	0.086
C_4	O_{127}	32	2	1	0	0	1	0.852	–0.004	0.181
C_5	O_{132}	32	1	1	0	0	1	0.841	–0.006	0.281
C_6	O_{133}	16	1	1	0	0	1	0.821	–0.011	0.489
C_7	O_{31}	64	4	1	1	0	0	0.649	–0.012	0.548
C_8	O_{151}	64	4	1	1	0	1	0.620	–0.013	0.632
C_9	O_{37}	32	2	1	1	0	0	0.582	–0.015	0.704
C_{10}	O_{61}	64	4	1	0	1	0	0.516	–0.015	0.621
C_{11}	O_{182}	32	4	1	0	1	1	0.460	–0.016	0.713
C_{12}	O_{187}	32	2	1	0	1	1	0.443	–0.021	0.912
C_{13}	O_{192}	32	1	1	0	1	1	0.435	–0.021	0.955
C_{14}	O_{188}	16	2	1	0	1	1	0.433	–0.025	1.019
C_{15}	O_{193}	16	1	1	0	1	1	0.425	–0.026	1.110

proposed system that uses these 15 configurations to dynamically adjust the encoding computational complexity in order to maintain the encoding time per GOP under a determined target.

7.2 Encoding Time Control System

The encoding time limitation algorithm presented in this chapter operates by changing the encoder operational point across the Pareto frontier found in the previous section in order to adjust the computational complexity, so that the encoding time per GOP remains below the upper bound. For this purpose, those configurations that appear in at least one of the two Pareto frontiers analysed in Sect. 7.1 were selected to be used by an encoding time control algorithm that is enabled only when the encoding time per GOP reaches values above the limit. In total, 15 configurations were selected and sorted in descending order of normalised computational complexity, as shown in Table 7.2.

The first column of Table 7.2 lists the ordered configurations names used in the system (C_k, k=1.15), while the second column presents the corresponding configuration index in the original 240 analysed cases (O_c, c=1.240). The *SR*, *BPR*, *HME*, *CTET*, *PUET* and *TUET* columns show the value of each parameter presented in Sect. 7.1 and the complexity, BD-PSNR and BD-rate columns present the results for each configuration in terms of normalised complexity, BD-PSNR and BD-rate, taking configuration 1 (C_1) as reference.

Notice that configurations C_2 and C_3 provide average complexity reductions of 5.5 and 11 %, respectively, with very small encoding performance loss. The remaining configurations (C_4–C_{15}) yield computational complexity reductions varying from 15 to 58 % with larger but still small encoding performance losses. In most cases, the complexity difference between two neighbouring configurations is under 7 %, which is acceptable for a time control of medium to large granularity. However, finer control granularity cannot be achieved by changing only the encoding configuration, so that some other finer adjustment procedure is necessary. Therefore, the system proposed in this section is made up of two cooperating control procedures: a medium-granularity encoding time control (MGTC), based on the configurations presented in Table 7.2, and a fine-granularity encoding time control (FGTC), based on the CCUPU method previously presented in Sect. 5.6.

Initially, the encoding process starts at full computational complexity, i.e. using the parameters defined in C_1. If the encoding time per GOP is below the limit, the control system stays idle and the encoding process is not constrained in any way, i.e. the encoder configuration used is C_1. Otherwise, if there is a need for any complexity saving to reduce the encoding time per GOP, MGTC is enabled, and the first configuration with reduced complexity (i.e. C_2) is chosen for the next GOP. If further complexity savings are necessary to reduce even more the encoding time, MGTC chooses a new configuration C_k that corresponds to the level of required complexity reduction to encode the next GOP. This process is repeated until the target encoding time per GOP is reached or until the granularity of the required complexity adjustment is finer than that achievable by MGTC. In this case, FGTC is enabled, and the encoding time control based on the CCUPU algorithm is executed by adjusting the number of constrained CTUs. Recall that in CCUPU two possible types of constrained CTUs are defined: those where exhaustive search is not performed in the CU decision process and those where exhaustive search is not performed in the PU decision process.

The proposed system is depicted in Fig. 7.4. The *complexity reduction factor computation* (CRFC) block is responsible for calculating the complexity adjustment factors required to maintain the encoding time per GOP close to the target. These factors are shown in Fig. 7.4 as $R_{\mathrm{T}(i)}$ and $\alpha^{\mathrm{GOP}}{}_{(i)}$, which are the medium and fine-granularity complexity adjustment factors, respectively, defined later in this section. Once the adjustment factors are known, the MGTC and FGTC blocks decide the best encoding configuration and parameters for medium and fine-granularity time control, respectively. Finally, the chosen configuration and parameters are used to reconfigure the HEVC encoder after each GOP, whenever necessary. The operations of the three blocks (CRFC, MGTC, FGTC) are explained in the next subsections. As the CRFC block calculates different complexity reduction factors for MGTC and FGTC, its operations are explained in the corresponding sections.

7.2.1 Medium-Granularity Time Control

The MGTC block shown in Fig. 7.4 is responsible for determining the encoder operational configuration based on the ratio $R_{\mathrm{T}(i)}$ between the target time (T_{T}) and the observed time ($T_{\mathrm{W}(i)}$), i.e. a weighted average of the encoding times of the last two

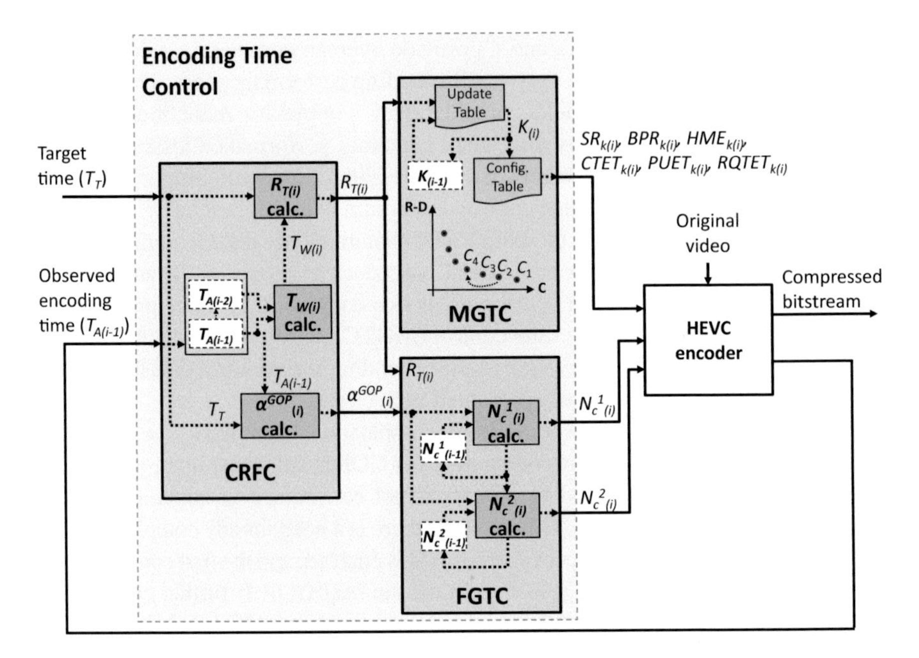

Fig. 7.4 Encoding time control system integrated with the HEVC encoder

GOPs. The calculation of $R_{\mathrm{T}(i)}$, defined in Eq. (7.1), is illustrated in the CRFC block of Fig. 7.4. The weighted average $T_{\mathrm{W}(i)}$ is calculated as given by Eq. (7.2), where $T_{\mathrm{A}(i-1)}$ and $T_{\mathrm{A}(i-2)}$ are the encoding times of GOPs $i-1$ and $i-2$ (i is the current GOP index). The weighting factors 2 and 1 applied to $T_{\mathrm{A}(i-1)}$ and $T_{\mathrm{A}(i-2)}$ in Eq. (7.2) were empirically found through simulations, as follows. In an initial system implementation, no weighting factors were used, and a simple average was computed between $T_{\mathrm{A}(i-1)}$ and $T_{\mathrm{A}(i-2)}$. Then, a second version with weights 2 and 1 for $T_{\mathrm{A}(i-1)}$ and $T_{\mathrm{A}(i-2)}$ and a third version with weights 3 and 1 were implemented. The average difference between T_{T} and $T_{\mathrm{W}(i)}$ was computed for the three versions, resulting in 3.7, 2.9 and 4.2, respectively. The smallest difference (2.9) defined the weights to be used in Eq. (7.2):

$$R_{\mathrm{T}(i)} = \frac{T_{\mathrm{T}}}{T_{\mathrm{W}(i)}} \tag{7.1}$$

$$T_{\mathrm{W}(i)} = \frac{2 \cdot T_{\mathrm{A}(i-1)} + 1 \cdot T_{\mathrm{A}(i-2)}}{3} \tag{7.2}$$

Once $R_{\mathrm{T}(i)}$ is computed, it is used to index a look-up table containing the precalculated ratios between the normalised complexities of the 15 configurations, to determine which one should be used to adjust the encoding time. The look-up table is shown as *Update Table* in Fig. 7.4 and is presented in Table C.2 of Appendix C, where line numbers indicate the current configuration, column numbers indicate the configuration for the next GOP and the values within each cell indicate the ratio between the normalised complexities of the next and the current GOP configurations. For example, the ratio between C_8 and C_4 is 0.728, which means that C_4 is 27.2 % less complex than C_8. Thus, if the encoder is operating in C_8 and $R_{T(i)}$ indicates that a complexity decrease around 27 % is required, C_4 must be selected as the encoding configuration for the next GOP. The new configuration index k is used to access a second look-up table (*Configuration Table*, in Fig. 7.4) to find the set of parameters ($SR_{k(i)}$, $BPR_{k(i)}$, $HME_{k(i)}$, $CTET_{k(i)}$, $PUET_{k(i)}$ and $RQTET_{k(i)}$) that will be used to encode the current GOP i. The *Configuration Table* is composed of columns 3–8 from Table 7.2.

Even though MGTC adjusts the encoding time by changing the operational configuration, a set of pre-evaluations revealed oscillations in the encoding time per GOP for certain target times. This happens because the controller cannot always find a configuration among the 15 in the *Update Table* that yields encoding times close enough to the target. In such cases, the encoder alternates between two configurations: one that yields encoding times under the target and another that yields encoding times above it. To attenuate this problem, a fine-granularity control was used, as explained in the next subsection.

7.2.2 Fine-Granularity Time Control

Section 5.6 of Chap. 5 presented the CCUPU method, which yielded the best complexity scaling accuracy among all methods proposed in that chapter, as well as the best R-D efficiency results. Now, this section describes how the CCUPU method was integrated with the encoding time control system presented in the previous section, aiming at providing a finer granularity level to the control algorithm and a wider range of achievable target times.

As shown in Fig. 7.4, the FGTC block receives as input the ${\alpha^{GOP}}_{(i)}$ parameter, which was calculated in the CRFC block. The ${\alpha^{GOP}}_{(i)}$ parameter is based on the α^{GOP} adjusting parameter used in CCUPU (see Sect. 5.6), which is the ratio between the encoding time of the previous GOP ($T_{A(i-1)}$) and the target time (T_T), as shown in Eq. (7.3). The value of ${\alpha^{GOP}}_{(i)}$ is then used to adjust the number ${N_c^k}_{(i)}$ of constrained CTUs per frame in the current GOP i, as shown in Eq. (7.4), where k indicates which of the two CCUPU complexity constraining parameters is used.[3] Once ${N_c^1}_{(i)}$ reaches its limit (i.e., the number of CTUs in a frame), the second parameter in the

[3] Recall that in the CCUPU method, N_c^1 represents the number of CTUs that allow using PUs smaller than $2N\times2N$ and N_c^2 is the number of CTUs with maximum coding tree depth constrained according to the depth used in spatially and temporally neighbouring CTUs.

scheme is used, represented as $N_{c(i)}^{2}$. Notice that when FGTC is applied, the encoding operating point chosen in the last execution of MGTC is used, so that the operating point parameters are still retrieved and used to configure the HEVC encoder:

$$\alpha_{(i)}^{\mathrm{GOP}} = \frac{T_{\mathrm{A}(i-1)}}{T_{\mathrm{T}}} \tag{7.3}$$

$$N_{\mathrm{c}(i)}^{k} = \alpha_{(i)}^{\mathrm{GOP}} \cdot N_{\mathrm{c}(i-1)}^{k} \tag{7.4}$$

7.2.3 *Encoding Time Limitation Algorithm*

The overall algorithm for the proposed encoding time limitation is presented in Fig. 7.5. From lines 01–12, it continuously monitors the running complexity and decides whether any encoding time reduction is needed. When an encoding time reduction is required (line 05), the control is activated and k is set to 2 (line 07), which means that configuration C_2 is used to encode the next GOP i. In line 13, the ratio $R_{\mathrm{T}(i)}$ is calculated and its value is used to trigger either the medium or the fine-granularity encoding time control. If $R_{\mathrm{T}(i)}$ shows that an adjustment larger than 15 % (either positive or negative) is required, the medium-granularity control takes place. Otherwise, the fine-granularity control starts.

For some video sequences, MGTC could still achieve acceptable results even when $R_{\mathrm{T}(i)}$ is under 15 %. However, since the system must be generic and yield accurate time control for most cases, the threshold was defined through experimental simulations aiming at the smallest average error between target and encoding times. The analysis presented in Sect. 7.1.2 shows that the largest gap in terms of computational complexity between two points in the Pareto frontiers occurs between configurations C_6 and C_7 (see Fig. 7.3 and Table 7.2). As this difference is around 17 %, a threshold of 17 % was initially used to enable FGTC in the first experiments. In subsequent experiments, this threshold was decreased to 16, 15 % and so on, until the average error between T_{T} and $T_{\mathrm{W}(i)}$ stopped decreasing, which happened when the threshold was set to 14 %.

7.3 Experimental Results

The performance of the encoding time limitation system was evaluated by encoding a set of video sequences using nine target times, defined as a percentage of the total encoding time used by an unconstrained HEVC encoder. In practical implementations of an HEVC encoder, the computational resources available in the encoding platforms found in the target devices will dictate the time constraints imposed on the video encoding procedure. In the experiments presented in this section, for test purposes only, the target times were defined with reference to an average time per

```
01  i ← 1, k ← 1, Nc^1 ← 0, Nc^2 ← 0
02  encode GOP i with SR_k(i), BPR_k(i), HME_k(i), CTET_k(i), PUET_k(i), RQTET_k(i)
03  T_A(i) ← time spent to encode GOP i
04  i ← i + 1
05  if(T_A(i-1) > T_T)
06     if(k = 1)
07        k ← 2
08        go to line 02
09     else
10        go to line 13
11  else
12     go to line 02
13  calculate R_T(i)
    ------------------------------------------------------------------ MGTC
14  if((R_T(i) > 1.15) OR (R_T(i) < 0.85))
15     k ← the configuration index indicated by R_T(i)
16     Nc^1_(i) ← 0, Nc^2_(i) ← 0
    ------------------------------------------------------------------ FGTC
17  else
18     calculate α^GOP_(i)
19     if(Nc^1_(i-1) < nCTU)
20        calculate Nc^1_(i)
21     else
22        calculate new Nc^2_(i)
23  for each i from 0 to nFR
24    sort CTUs in ascending order of R-D cost
25    for each j from 0 to nCTU
26      if(j < Nc^k)
27        mark CTU j as constrained
28      else
29        mark CTU j as unconstrained
30      encode CTU j with SR_k(i), BPR_k(i), HME_k(i), CTET_k(i), PUET_k(i), RQTET_k(i)
31    if last frame go to line 01
    ------------------------------------------------------------------
32  T_A(i) ← time spent to encode GOP_(i)
33  i ← i + 1
34  go to line 13
```

Fig. 7.5 Pseudocode for the encoding time limitation system

GOP computed for the following six high-resolution (1080p) sequences, all of which are described in Appendix A: *Beauty*, *Bosphorus*, *HoneyBee*, *Jockey*, *ShakeNDry* and *YachtRide*. The first 150 frames of each sequence were encoded with HM (version 13) [1], QPs 22, 27, 32 and 37, and the *Main*, *low delay* configuration [2]. Based on the average encoding time per GOP (372 s), nine target times were defined: 37 s (10 % of 372 s), 74 s (20 %), 111 s (30 %), 149 s (40 %), 186 s (50 %), 223 s (60 %), 260 s (70 %), 298 s (80 %) and 335 s (90 %).

Table 7.3 shows average results in terms of encoding time limitation accuracy and R-D efficiency for the nine target times tested in the experiments. Encoding time limitation accuracy was measured as inversely proportional to the encoding

Table 7.3 Average encoding time error (TE), BD-rate and BD-PSNR results

Target (s/GOP)	Average TE (%)	BD-rate (%)	BD-PSNR (dB)	Enc. time reduction (%)
335	4.704	0.160	–0.005	10
298	6.373	0.352	–0.011	20
260	2.448	0.505	–0.016	30
223	1.405	0.742	–0.023	40
186	1.676	0.972	–0.028	50
149	3.417	1.193	–0.035	60
111	2.556	1.709	–0.048	70
74	0.736	3.880	–0.111	80
37	2.691	9.839	–0.288	90

time error (TE), which is calculated as the average absolute difference between encoding time and target time, considering all GOPs after the settling phase of the control. The settling phase usually lasts for about ten GOPs, as shown later in this section. The second column of Table 7.3 shows that encoding time errors varied from 0.736 to 6.373 % (averaging 2.889 %), which means that the algorithm is quite accurate, since it is capable of constraining the encoding times without significant deviations from the target.

The third and fourth columns of Table 7.3 show average results in terms of compression efficiency for each target time. Notice that BD-rate and BD-PSNR values tend to increase and decrease, respectively, as the target time reduces. This is expected because tighter target times demands that less complex (which are usually less efficient) encoding configurations are chosen by MGTC and that a larger number of constrained CTUs are set by FGTC. The table shows that for target times between 335 s and 111 s, very small compression efficiency losses are noticed, but these losses increase significantly when the target time is reduced to 74 s and 37 s. Still, considering the very large encoding time reduction in comparison to the original unconstrained encoder (80 and 90 %, respectively), the losses are quite acceptable. It is important to highlight that the encoding time could be reduced by up to 90 % only by combining MGTC and FGTC, a reduction not possible to achieve by the complexity reduction methods presented in Chap. 6 or by the CCUPU algorithm presented in Sect. 5.6 separately. In fact, no other work found in the literature so far reported performance results for time reductions near 90 % for the HEVC encoding process, as shown later in Sect. 7.4.

Figures 7.6 and 7.7 show the operation of the proposed encoding time limitation system for all frames (150 GOPs) of two video sequences (*Beauty* and *YachtRide*, respectively). Figure 7.6a shows encoding times per GOP, Fig. 7.6b shows the evolution of the encoding configuration index *k*, Fig. 7.6c shows the evolution of the N_c^1 parameter and Fig. 7.6d shows the evolution of the N_c^2 parameter when encoding the *Beauty* sequence. The same applies to Fig. 7.7, with respect to the *YachtRide* sequence. The lines in each chart correspond to the nine different encoding times tested. Hidden lines due to overlapping are identified with labels at the right side of the charts.

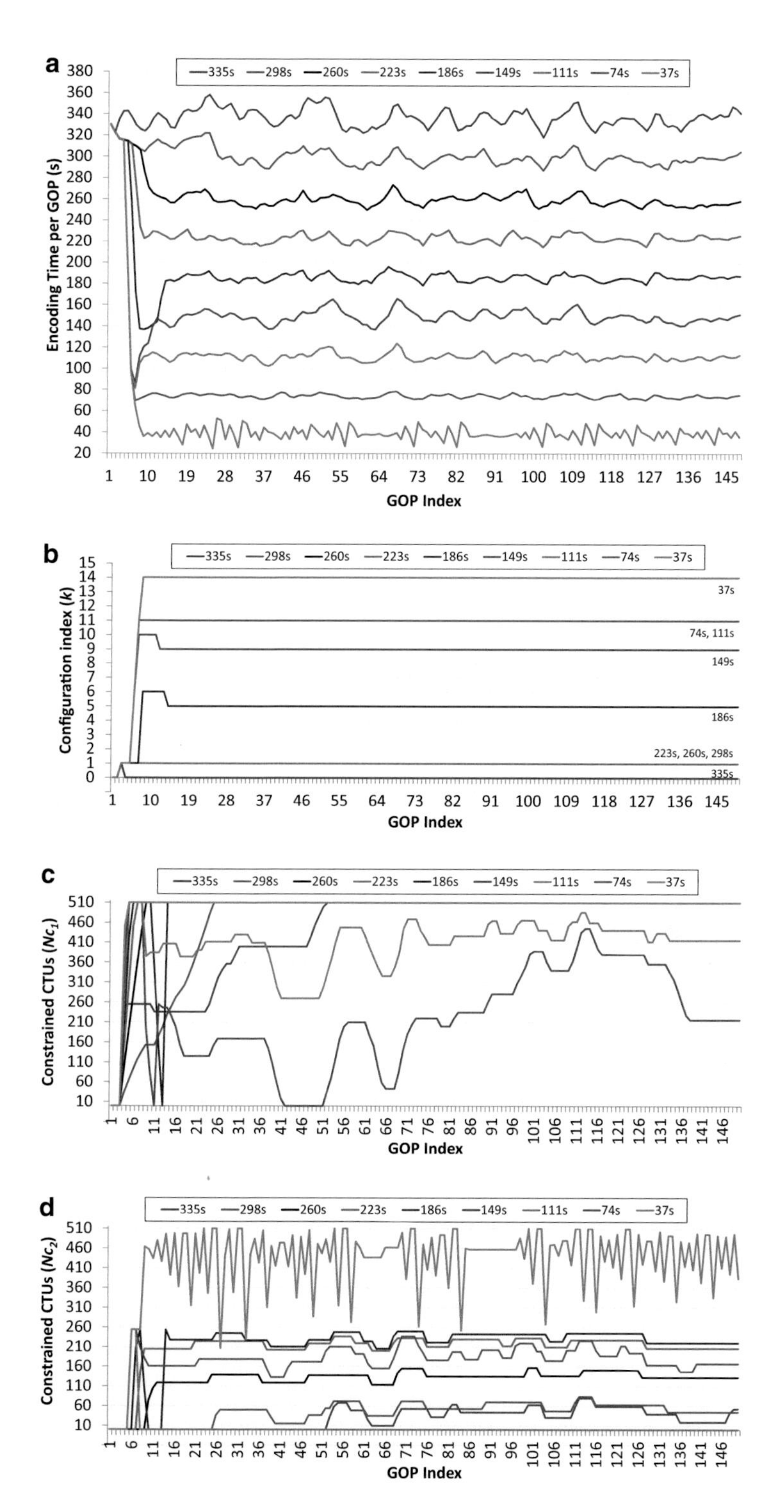

Fig. 7.6 (**a**) Encoding times, (**b**) configuration index k, (**c**) $N^1_{c(i)}$ parameter and (**d**) $N^2_{c(i)}$ parameter for each GOP in the *Beauty* sequence (QP 22)

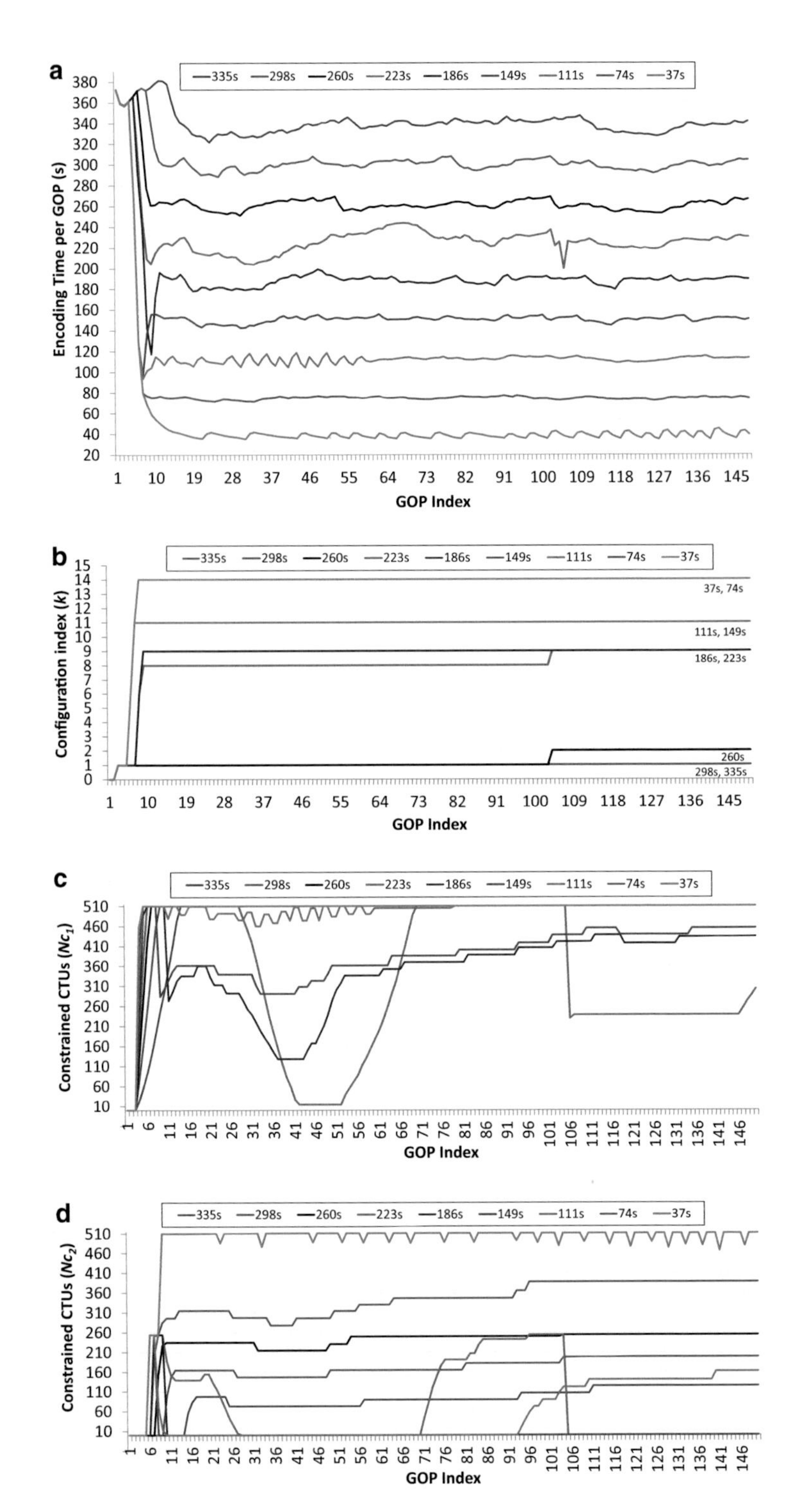

Fig. 7.7 (**a**) Encoding times, (**b**) configuration index k, (**c**) $N^1_{c(i)}$ parameter and (**d**) $N^2_{c(i)}$ parameter for each GOP in the *YachtRide* sequence (QP 32)

Notice in Figs. 7.6a and 7.7a that the encoding times settle around the target by the tenth GOP. This is the time taken by MGTC to find the best encoding configuration in the *Update Table* and by FGTC to determine the amount of constrained CTUs so that the target time is achieved. Notice in Figs. 7.6b and 7.7b that the encoding configuration indexes are larger when small target times are set (i.e. low-complexity configurations are chosen), but there is little or no change in the index once it settles. This happens because MGTC is only enabled when a complexity adjustment larger than 15 % is required. In the remaining cases, FGTC performs the adjustments, and this is why the parameters in Figs. 7.6c, d and 7.7c, d vary much more. The behaviour of the proposed method when encoding the four remaining test sequences, *Bosphorus*, *HoneyBee*, *Jockey*, *ShakeNDry*, was similar to that just described.

In the experiments described in this section, the most complex sequence tested was the *YachtRide* (Fig. 7.7). Notice that in this case, when considering the smallest encoding target time (37 s), all control parameters (configuration index k, N_c^1 and N_c^2) saturate in most GOPs, but the target time is still achieved with small errors. If the complexity was set to a value smaller than 37 s, the control parameters would still saturate in the same manner, but the target encoding time would not be achieved.

To evaluate the performance of the proposed system in the presence of sudden changes of the video content characteristics (spatial and temporal activities), several composite sequences made up by concatenating two different sequences were encoded. In Fig. 7.8, a graph is shown presenting the encoding time per GOP for a video sequence created by concatenating the two sequences *Beauty* and *YachtRide*, previously analysed separately in Figs. 7.6 and 7.7. In the concatenated sequence, the scene change occurs at GOP 150, close to the midpoint of the chart. Notice that some oscillations appear after the scene change due to the settling phase of the algorithm. This settling phase also appears in GOPs 1–10, but is much shorter at the scene change because the encoder is already using a complexity reduction configuration. In the specific case of target time 186 s, the oscillations are more prominent because there are less points in the Pareto frontier near configurations C_6 and C_7 (see the gap in the Pareto frontiers of Fig. 7.3 between complexities 0.64 and 0.82). We can conclude that the control method proposed is able to rapidly accommodate to changes in video content characteristics.

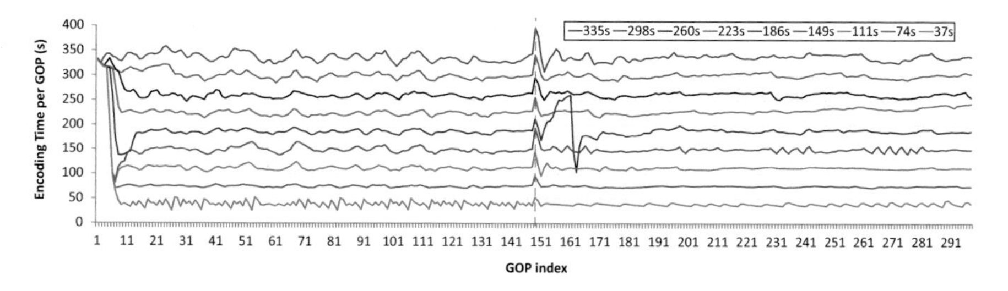

Fig. 7.8 Encoding times for each GOP in the concatenation of the *Beauty* and *YachtRide* sequences (QP 32), with the scene change in GOP 150

7.4 Comparison with Previous Works

Table 7.4 presents a comparison between the proposed encoding time control and strategies found in related works [3–18]. It is important to highlight that all works found in the literature are actually complexity reduction and scaling methods that apply different strategies to reduce the overall complexity of the HEVC encoding process (e.g. for a complete sequence). This way, they cannot be directly compared with the system presented in this chapter in terms of encoding time per GOP and deviation from the target time. Still, this section presents a comparison of coding efficiency considering average encoding time reductions, since these are the values made available in those works.

Only works that provide results in terms of BD-rate increase, followed the CTC settings and used the *Main* encoder configuration are compared in Table 7.4. Works [3–5] report results obtained with the *all intra* temporal configuration, works [6–10] report results for the *low delay* configuration and works [11–18] report results for the *random access* configuration. The second column of Table 7.4 presents the average encoding time reduction achieved by each work, while the third column presents the associated average BD-rate increase, in comparison to the unmodified HM encoder. Notice that works [8, 11, 15] describe complexity scaling strategies, so that their results in Table 7.4 represent the achievable range of encoding time reduction.

The table shows that the system presented in this chapter excels both in terms of average encoding time reduction and compression efficiency. The only method that achieves similar reduction levels is [15], which allows decreasing the encoding

Table 7.4 Comparison with related works

Reference	Enc. time reduction (%)	BD-rate (%)
Seunghyun [3]	50	0.6
Khan [4]	44	1.3
Zhang [5]	60	1.0
J.-Hyeok [6]	48	1.2
Goswami [7]	38	1.7
Zhao [8]	23–53	0.2–2.4
Shi [9]	22	1.4
Correa [10]	65	1.3
Vanne [11]	32–50	0.3–1.3
Liquan [13]	41	1.15
Shen [14]	38	1.9
Correa [15]	9–79	0.02–14.5
Ahn [16]	50	1.4
Lee [17]	69	3.0
Shen [18]	49	0.7
Wei-Jhe [19]	43	5.1
Proposed	**10–90**	**0.16–9.84**

times in up to 79 % with an average BD-rate increase of 14.5 %. As previously shown, the system presented in this chapter achieves encoding time reductions of up to 90 %, with average BD-rate increases of 9.84 % in that case, which means that it is significantly more efficient than [15]. The remaining works in Table 7.4 achieved average encoding time reductions between 23 and 69 %, with average BD-rate increases between 0.6 and 2.4 %. Notice that, besides better R-D performance, the control system presented in this chapter is able to guarantee an upper bound for the encoding time, whereas the related works do not have such capability.

7.5 Conclusions

The encoding time limitation system for HEVC presented in this chapter was developed based on an R-D-C analysis of a set of encoder operational configurations, all of which have been created based on the complexity reduction and scaling algorithms previously presented in this book. The system integrates a medium-granularity time control and a fine-granularity time control to allow adjusting the encoder operation, aiming at maintaining the encoding time per GOP under an upper bound. Experimental results have shown that the method accurately limits the encoding time per GOP, with an average deviation from the target of only 2.9 %. Encoding time reduction levels of up to 90 % can be achieved (in comparison to the unmodified HEVC encoder), with compression efficiency losses that depend on the target time. Target times that required small reductions in complexity were achieved with negligible losses (BD-rate increases under 1 %), while target times that resulted into larger complexity reductions implied small to medium losses (BD-rate increases between 1.2 and 9.8 %).

The method presented in this chapter is especially useful in devices where computing resources, such as CPUs, must be shared among several applications, requiring an adjustable time limitation procedure. In addition, the complexity reduction levels achieved show that the method can also contribute to extend the battery life of autonomous devices that rely on software HEVC video encoders.

References

1. ISO/IEC-JCT1/SC29/WG11, High Efficiency Video Coding (HEVC) Test Model 13 (HM 13) Encoder Description, Geneva, Switzerland (2013)
2. ISO/IEC-JCT1/SC29/WG11, Common test conditions and software reference configurations, Geneva, Switzerland (2012)
3. C. Seunghyun, K. Munchurl, Fast CU splitting and pruning for suboptimal CU partitioning in HEVC intra coding. Circuits Syst Video Technol IEEE Trans **23**, 1555–1564 (2013)
4. M.U.K. Khan, M. Shafique, J. Henkel, An Adaptive Complexity Reduction Scheme with Fast Prediction Unit Decision for HEVC Intra Encoding, in *IEEE International Conference on Image Processing (ICIP)* (2013)

5. Z. Hao, M. Zhan, Fast intra mode decision for high efficiency video coding (HEVC). Circuits Syst Video Technol IEEE Trans **24**, 660–668 (2014)
6. L. Jong-Hyeok, P. Chan-Seob, K. Byung-Gyu, Fast coding algorithm based on adaptive coding depth range selection for HEVC, in *Consumer Electronics—Berlin (ICCE-Berlin), 2012 IEEE International Conference on* (2012), pp. 31–33
7. Kalyan Goswami, Byung-Gyu Kim, Dong-San Jun, Soon-Heung Jung, J.S. Choi, Early Coding Unit (CU) Splitting Termination Algorithm for High Efficiency Video Coding (HEVC), *to be published in ETRI Journal* (2014)
8. T. Zhao, Z. Wang, S. Kwong, Flexible mode selection and complexity allocation in high efficiency video coding. Selected Topics Signal Process IEEE J **7**, 1135–1144 (2013)
9. Y. Shi, Z. Gao, X. Zhang, Early TU Split Termination in HEVC Based on Quasi-Zero-Block, in *3rd International Conference on Electric and Electronics* (2013)
10. G. Correa, P. Assuncao, L. Agostini, L.A. da Silva Cruz, Fast hevc encoding decisions using data mining. IEEE Trans Circuits Syst Video Technol **24**, 660–673 (2015)
11. J. Vanne, M. Viitanen, T. Hamalainen, Efficient mode decision schemes for HEVC inter prediction. Circuits Syst Video Technol IEEE Trans **24**, 1579–1593 (2014)
12. H. Wei-Jhe, H. Hsueh-Ming, Fast coding unit decision algorithm for HEVC, in *Signal and Information Processing Association Annual Summit and Conference (APSIPA), 2013 Asia-Pacific* (2013), pp. 1–5
13. S. Liquan, L. Zhi, Z. Xinpeng, Z. Wenqiang, Z. Zhaoyang, An effective CU size decision method for HEVC encoders. Multimedia IEEE Trans **15**, 465–470 (2013)
14. S. Xiaolin, Y. Lu, and C. Jie, Fast coding unit size selection for HEVC based on Bayesian decision rule, in *2012 Picture Coding Symposium* (2012), pp. 453–456
15. G. Correa, P. Assuncao, L. Agostini, L. Silva Cruz, Complexity scalability for real-time HEVC encoders, J *Real-Time Image Process,* pp. 1–16 (2014) http://link.springer.com/article/10.1007/s11554-013-0392-8
16. A. Sangsoo, L. Bumshik, K. Munchurl, A novel fast CU encoding scheme based on spatiotemporal encoding parameters for HEVC inter coding. Circuits Syst Video Technol IEEE Trans **25**, 422–435 (2015)
17. L. Jaeho, K. Seongwan, L. Kyungmin, L. Sangyoun, A Fast CU size decision algorithm for HEVC. Circuits Syst Video Technol IEEE Trans **25**, 411–421 (2015)
18. S. Liquan, Z. Zhaoyang, L. Zhi, Adaptive inter-mode decision for HEVC jointly utilizing inter-level and spatiotemporal correlations. Circuits Syst Video Technol IEEE Trans **24**, 1709–1722 (2014)
19. C. Kiho, E.S. Jang, Early TU decision method for fast video encoding in high efficiency video coding. Electronics Lett **48**, 689–691 (2012)

Chapter 8
Conclusions

This book presented a review of the main works focusing on computational complexity of HEVC encoders and a number of contributions with novel solutions for reducing, scaling and controlling the computational complexity. The research also included extensive simulations and experiments focused on the study and critical analysis of computational complexity for HEVC. In this chapter, the book is concluded with a summary of the main achievements, as well as some directions for future research.

8.1 Final Remarks

The HEVC standardisation process, which started in early 2010 and was finalised in 2013, incorporated a series of encoding tools and functionalities to the basic hybrid video compression system used in the most recent previous standards. By employing such tools and functionalities, HEVC is able to achieve average bit rate reductions of 40–50 % in comparison with H.264/AVC, at the same subjective image quality. However, as Chap. 4 has shown, these improvements came with associated increases in the encoder computational complexity.

The existence of few works focusing on computational complexity analysis, reduction and scaling for HEVC motivated the study and the analysis of the encoder compression efficiency and computational complexity, as presented in Chap. 4, in order to identify which are the most computationally demanding operations in the encoding process. Based on the results of such analysis, a set of methods for complexity reduction and scaling were proposed in Chaps. 5, 6 and 7.

G. Corrêa et al., *Complexity-Aware High Efficiency Video Coding*,
DOI 10.1007/978-3-319-25778-5_8

8.2 Research Contributions

Following the state-of-the-art review presented in Chap. 3, the first main contribution of this book is presented in Chap. 4 and consists of an experimental investigation and analysis of the R-D efficiency and computational complexity of HEVC encoders. Two separate analyses were performed to evaluate the encoding tools and the frame partitioning structures introduced by the HEVC standard. A set of encoding configurations was created to investigate the impact of each tool or frame partitioning structure, varying the encoding parameter set and comparing the results with a baseline encoder.

The results obtained in Chap. 4 allowed concluding that HEVC complexity can be largely decreased at practically no coding efficiency cost, if the coding tools are wisely combined and configured. It was observed that by first enabling those tools which provide higher R-D efficiency gains for the least computational complexity costs, a near-optimal trade-off between computational complexity and encoding efficiency can be achieved. An encoder configuration with high R-D efficiency and low computational complexity levels was proposed. Besides, a set of configurations which resulted in the largest computational complexity increases were identified for use in the R-D-C optimised scheme presented in Chap. 7. Still in Chap. 4, it was also found that the encoding computational complexity can be thoroughly reduced by managing the frame partitioning structures of HEVC, even though some configurations incur in much larger losses in R-D efficiency than others. The frame partitioning structures that produce the largest impact in the computational complexity and the best trade-off between complexity and R-D efficiency were selected for the development of the complexity reduction and scaling strategies presented in Chaps. 5 and 6.

The complexity scaling algorithms proposed in Chap. 5 represent the second main contribution of this book. All developed methods aim at dynamically adjusting the frame partitioning structures (CTUs, CUs and PUs) in order to adapt the encoding process according to the available computational complexity in the encoder. Five algorithms were proposed in this chapter (FDCR, VDCR, MCTDL, CTDE and CCUPU), and each one overcame the previous in terms of R-D efficiency for the same target complexities, which is an outcome obtained by adding up more intelligent ways of constraining the frame partitioning structures of HEVC. In general, the complexity scaling accuracy of the five algorithms is quite similar, and all of them achieve running complexities very close to the targets, even though the last method (CCUPU) provides much larger computational complexity reduction levels than the previous ones. While FDCR, VDCR, MCTDL and CTDE provide computational complexity reductions of up to 40 %, the CCUPU method is able to reduce it in up to 80 %. In terms of R-D efficiency, CCUPU operates in two regions: a near-lossless encoding, where complexity reductions of up to 50 % are achieved at the cost of small BD-rate increases (from 0.03 to 1.28 %), and a lossy region, where a complexity reduction of up to 80 % is achieved with a BD-rate increase between 3.98 % and 22.64 %. From the results obtained in Chap. 5, it was possible to notice that adding up information related to spatial and temporal correlation to the last complexity scaling algorithms decreased the R-D efficiency losses noticed in the first ones. This lead to the conclusion that if more information from the original video, as well as intermediate information computed during the encoding process, were taken into account in the development of complexity reduction and scaling methods, R-D efficiency losses even smaller than those observed with CCUPU could be achieved.

The third major contribution of this book was presented in Chap. 6. A set of classification trees obtained through data mining techniques was developed and implemented in the HM encoder to early terminate the exhaustive search for the best frame partitioning structure configuration. The three sets of decision trees were created to make use of intermediate encoding results for early terminating the determination of coding trees, PUs and RQTs. They were separately and jointly implemented to provide further complexity reduction levels. On average, a complexity reduction of 65 % was achieved when the three early terminations are jointly implemented, with a BD-rate increase of only 1.36 %.

However, the early termination methods proposed in Chap. 6 provide fixed complexity reductions, differently from the dynamic complexity scaling methods presented in Chap. 5. Besides, they do not guarantee that the encoding process is performed within a determined time budget. In order to solve this issue, Chap. 7 presents the fourth contribution of this book: an encoding time control system that combines the findings of Chap. 4, the best-performing complexity scaling method of Chap. 5 and the complexity reduction algorithms of Chap. 6, aiming at adjusting the encoder operating point whenever necessary so that the encoding time per GOP is kept under a specified target.

8.3 Future Work

Several research directions can be thought for the future, departing from what is presented in this book. One of the most important and immediate works is the implementation of the best methods found in the literature, as well as those presented as contributions in this book (e.g. the CCUPU algorithm, the decision trees and the encoding time control system), in optimised encoders developed for real-time applications. All the algorithms found in the technical literature and all strategies presented in this book are developed and tested on top of the HM encoder, which was developed during the standardisation process for testing purposes only. This way, even though similar complexity reduction rates are expected in optimised encoder implementations, investigations are required to assess the effect of applying such solutions.

Another trend to be explored in future work is the exploration of parallel computing strategies aiming at the achievement of real-time video coding. It is now common to find handheld devices equipped with multicore general processors and graphics processing units (GPUs), which can be used to increase encoding speed through the divide-and-conquer approach. As explained in Chap. 2, the HEVC standard includes a set of parallel processing structures (Dependent Slice Segments, Tiles, WPP) which can be explored to solve such issues. The contributions presented in this book, mostly based on the management of the HEVC frame partitioning structures, can be extended in future works to support the parallel processing structures, so that computational complexity could be reduced or scaled separately in each core.

All the analyses, algorithms and methods presented in this book focused on the encoding process, which is the most critical issue in terms of computational complexity in HEVC-based codec systems. However, the use of computational resources on the decoder side will also become more important with the introduction of higher spatial resolution, such as ultra-HD. Techniques for computational complexity reduction and scaling on video decoders will need to be further investigated, especially for devices with fewer computational capabilities. Furthermore, the heterogeneity of mobile devices, varying from those with fast multicore processors and GPUs to those with slower single-core processors, will also call for methods that allow scaling computational complexity on the decoder side according to the target platform constraints.

Appendix A
Common Test Conditions and Video Sequences

This appendix describes the main characteristics of the video sequences used in the experiments presented throughout this book. Section A.2 shows one frame belonging to each video sequence listed in Sect. A.1.

Video Sequences Characteristics

Besides the 24 video sequences listed in the CTC of JCT-VC [1], 11 supplementary video sequences were used in the experiments in order to allow tests with sequences not used in the training of the decision trees presented in Chap. 6 and in the parameter selection analysis presented in Sect. 7.1. The 11 supplementary sequences were obtained from the CTC of the Joint Collaborative Team on 3D Video Coding (JCT-3V) [2], from the Ultra Video Group at the Tampere University of Technology [3] and from the Multimedia Group at the Poznan University [4].

Table A.1 lists the 35 video sequences used in the experiments described in this book and presents their main characteristics. The rightmost column indicates the source of the video sequences, where CTC 2D stands for the CTC of JCT-VC, CTC 3D stands for the CTC of the JCT-3V, UVG stands for the Ultra Video Group at the Tampere University of Technology and POZ stands for Multimedia Group at the Poznan University. In the case of videos from JCT-3V, only the central views of multiview video sequences are used in the experiments.

Video Sequences

Trying to illustrate the characteristics of the 35 video sequences listed in Table A.1, the frame positioned exactly in the middle of each one is presented in this section. Figures A.1, A.2, A.3, A.4, A.5, A.6, A.7, A.8, A.9, A.10, A.11, A.12, A.13, A.14,

G. Corrêa et al., *Complexity-Aware High Efficiency Video Coding*,
DOI 10.1007/978-3-319-25778-5

Table A.1 Video sequence spatial resolutions

Name	Spatial resolution	Frame count	Frame rate (fps)	Bit depth	Source
BaskeballDrillText	832×480	500	50	8	CTC 2D
BasketballDrill	832×480	500	50	8	CTC 2D
BasketballDrive	1920×1080	500	50	8	CTC 2D
BasketballPass	416×240	500	50	8	CTC 2D
Beauty	1920×1080	600	120	8	UVG
BlowingBubbles	416×240	500	50	8	CTC 2D
Bosphorus	1920×1080	600	120	8	UVG
BQMall	832×480	600	60	8	CTC 2D
BQSquare	416×240	600	60	8	CTC 2D
BQTerrace	1920×1080	600	60	8	CTC 2D
Cactus	1920×1080	500	50	8	CTC 2D
ChinaSpeed	1024×768	500	30	8	CTC 2D
FourPeople	1280×720	600	60	8	CTC 2D
HoneyBee	1920×1080	600	120	8	UVG
Jockey	1920×1080	600	120	8	UVG
Johnny	1280×720	600	60	8	CTC 2D
Kimono	1920×1080	240	24	8	CTC 2D
KristenAndSara	1280×720	600	60	8	CTC 2D
NebutaFestival	2560×1600	300	60	10	CTC 2D
ParkScene	1920×1080	240	24	8	CTC 2D
PartyScene	832×480	500	50	8	CTC 2D
PeopleOnStreet	2560×1600	150	30	8	CTC 2D
Poznan_CarPark	1920×1080	600	25	8	POZ
Poznan_Hall1	1920×1080	200	25	8	POZ
Poznan_Street	1920×1080	250	25	8	CTC 3D
RaceHorses1	416×240	300	30	8	CTC 2D
RaceHorses2	832×480	300	30	8	CTC 2D
ShakeNDry	1920×1080	600	120	8	UVG
Shark	1920×1080	300	50	8	CTC 3D
SlideEditing	1280×720	300	30	8	CTC 2D
SlideShow	1280×720	500	20	8	CTC 2D
SteamLocomotive	2560×1600	300	60	10	CTC 2D
Tennis	1920×1080	150	30	8	CTC 2D
Traffic	2560×1600	150	30	8	CTC 2D
YachtRide	1920×1080	600	120	8	UVG

A.15, A.16, A.17, A.18, A.19, A.20, A.21, A.22, A.23, A.24, A.25, A.26, A.27, A.28, A.29, A.30, A.31, A.32, A.33, A.34 and A.35 show each middle frame, which are all pictured here in the same size despite their original resolution.

Fig. A.1 BaskeballDrillText

Fig. A.2 BaskeballDrill

Fig. A.3 BasketballDrive

Fig. A.4 BasketballPass

Fig. A.5 Beauty

Fig. A.6 BlowingBubbles

Fig. A.7 Bosphorus

Fig. A.8 BQMall

Fig. A.9 BQSquare

Fig. A.10 BQTerrace

Fig. A.11 Cactus

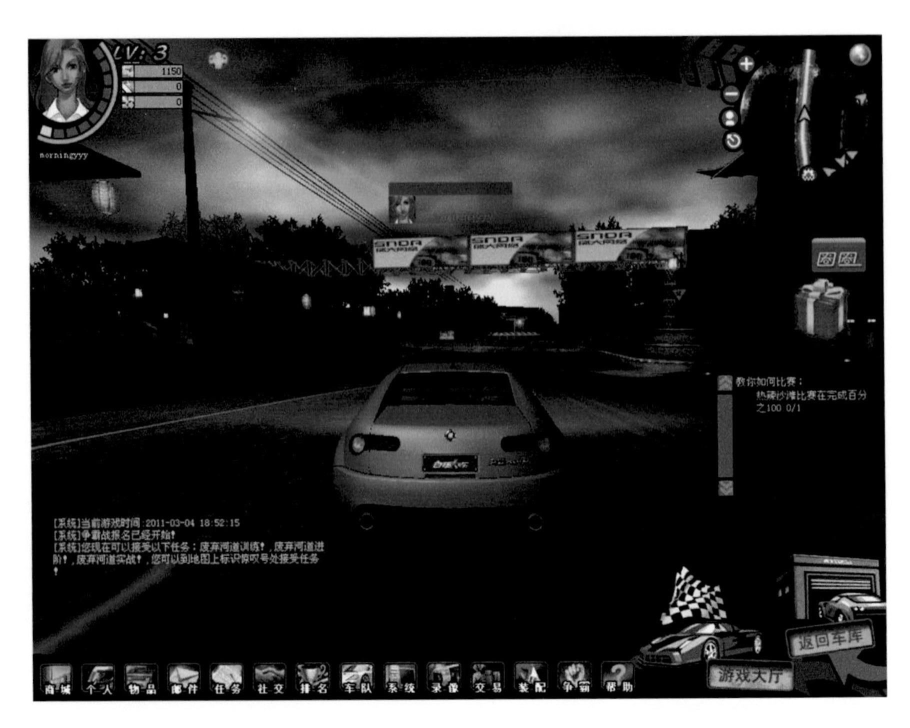

Fig. A.12 ChinaSpeed

Fig. A.13 FourPeople

Fig. A.14 HoneyBee

Fig. A.15 Jockey

Fig. A.16 Johnny

Fig. A.17 Kimono

Fig. A.18 KristenAndSara

Fig. A.19 NebutaFestival

Fig. A.20 ParkScene

Fig. A.21 PartyScene

Fig. A.22 PeopleOnStreet

Fig. A.23 Poznan_CarPark

Fig. A.24 Poznan_Hall1

Fig. A.25 PoznanStreet

Fig. A.26 RaceHorses1

Fig. A.27 RaceHorses2

Fig. A.28 ShakeNDry

Fig. A.29 Shark

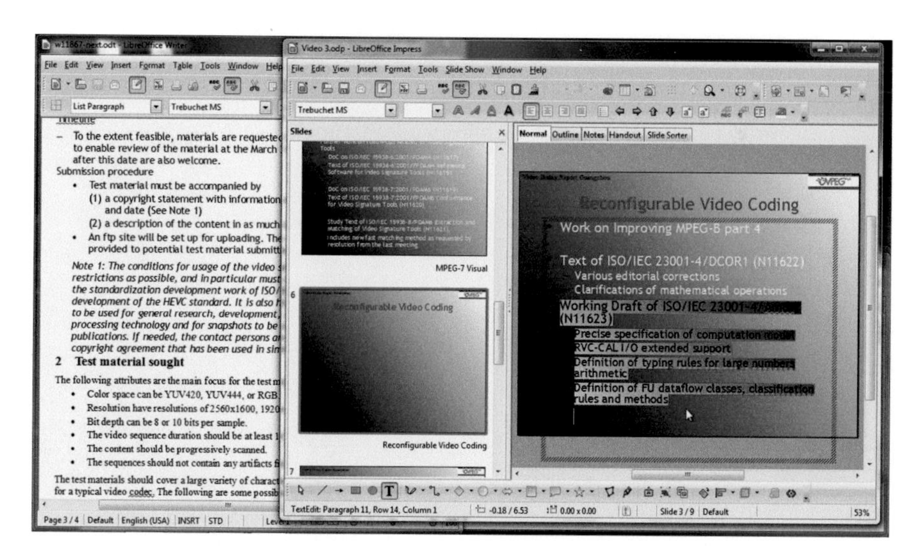

Fig. A.30 SlideEditing

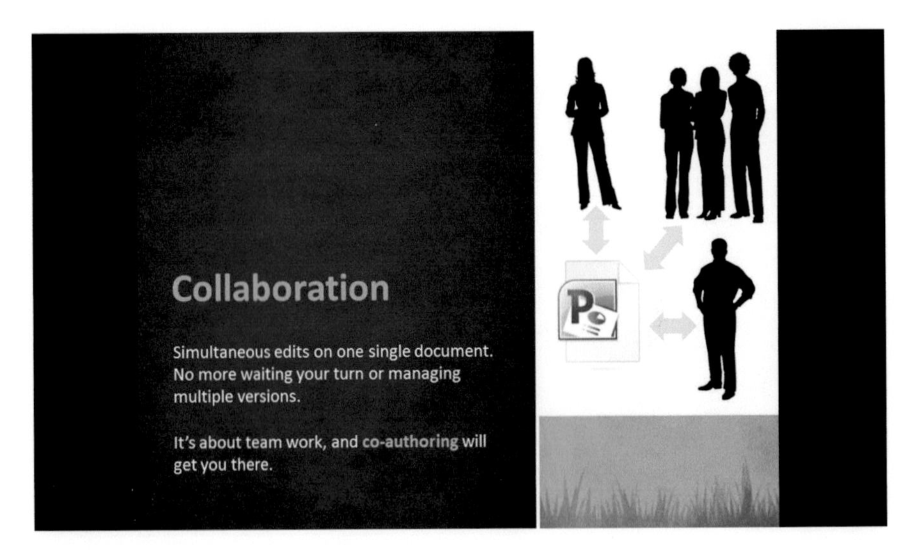

Fig. A.31 SlideShow

Fig. A.32 SteamLocomotive

Fig. A.33 Tennis

Fig. A.34 Traffic

Fig. A.35 YachtRide

References

1. ISO/IEC-JCT1/SC29/WG11, Common Test Conditions and Software Reference Configurations, Geneva, Switzerland, 2012
2. ISO/IEC-JCT1/SC29/WG11, Common Test Conditions of 3DV Core Experiments, San Jose, US, 2014
3. *Tampere University of Technology—Ultra Video Group*. Available: http://ultra-video.cs.tut.fi/
4. *Poznan University of Technology*. Available: http://www3.put.poznan.pl/

Appendix B
Obtained Decision Trees

This appendix presents the decision trees obtained with the methodology described in Chap. 6. The graphic representation of each tree, obtained with the WEKA tool [1], is presented in Sects. B.1, B.2 and B.3 for the coding tree early termination, the PU early termination and the RQT early termination, respectively.

Decision Trees for Coding Tree Early Termination

As explained in Sect. 6.3, three decision trees were trained and implemented for the coding tree early termination, one for each CU size that allows splitting into smaller CUs (i.e. 16×16, 32×32 and 64×64). The three trees are presented in Figs. B.1, B.2, and B.3, where *C* and *T* correspond to the decisions of continuing and terminating the CU splitting process, respectively.

Decision Trees for PU Early Termination

The four decision trees introduced in Sect. 6.4 for the PU early termination are presented in Figs. B.4, B.5, B.6 and B.7, one for each CU size possible (i.e. 8×8, 16×16, 32×32 and 64×64). In the figures, *C* and *T* correspond to the decisions of continuing and terminating the process of choosing the best PU splitting mode, respectively.

G. Corrêa et al., *Complexity-Aware High Efficiency Video Coding*,
DOI 10.1007/978-3-319-25778-5

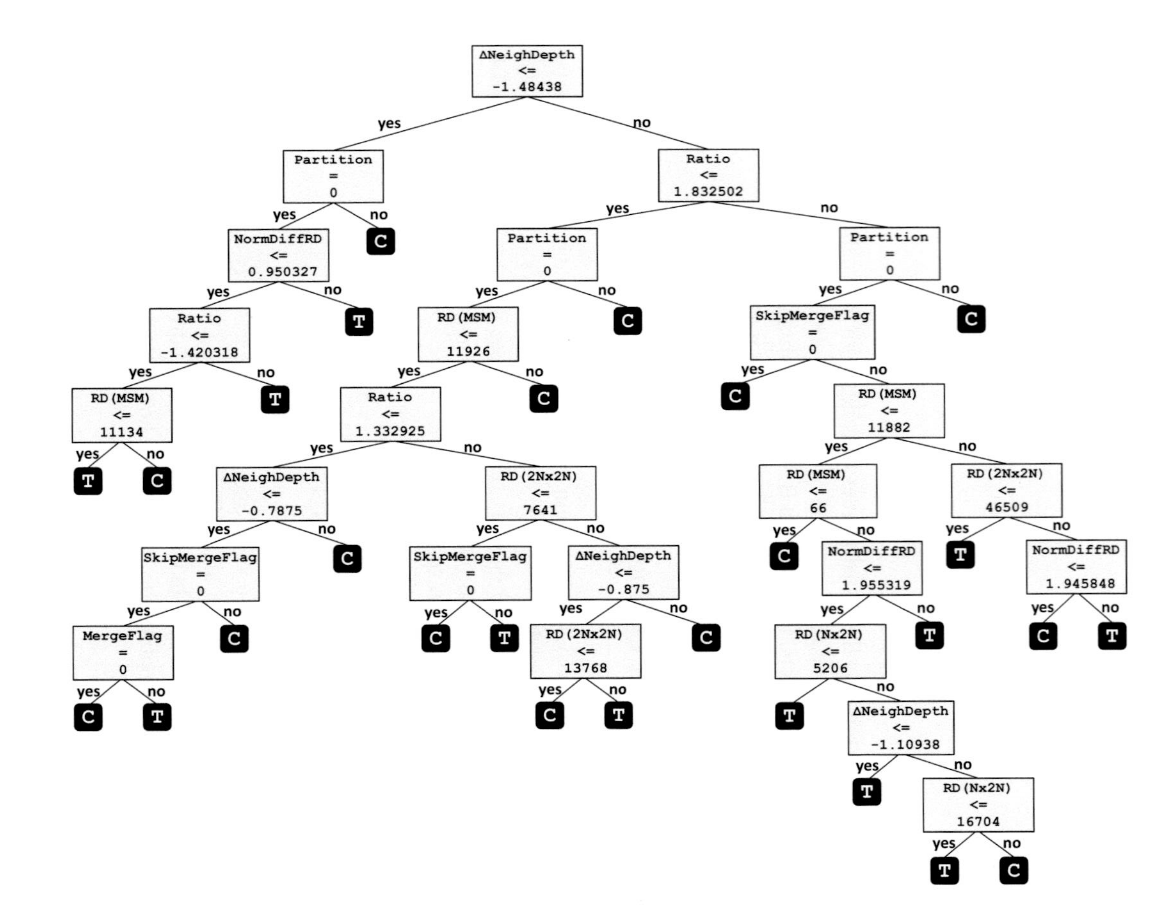

Fig. B.1 Coding tree early termination decision tree for 16×16 CUs

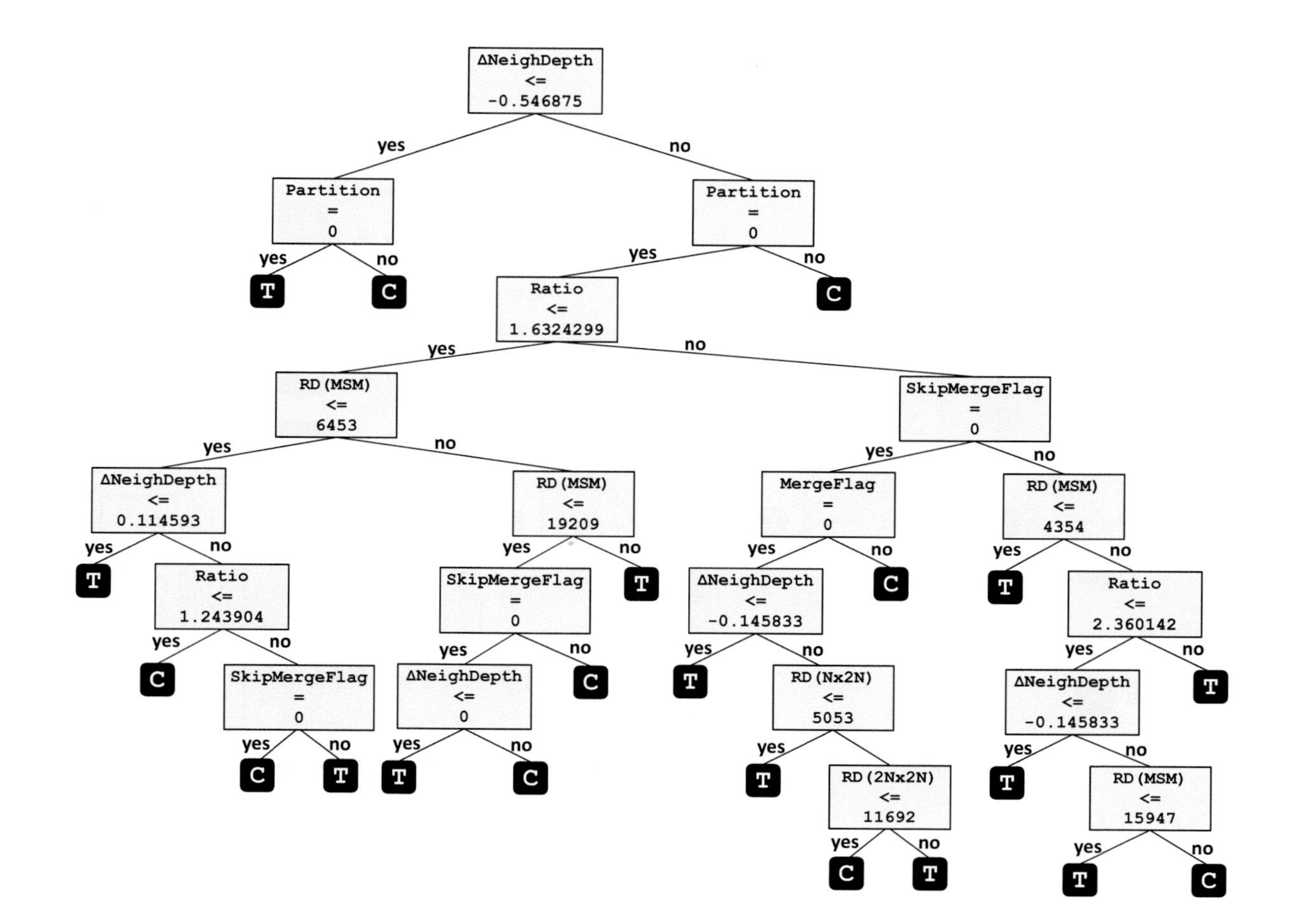

Fig. B.2 Coding tree early termination decision tree for 32×32 CUs

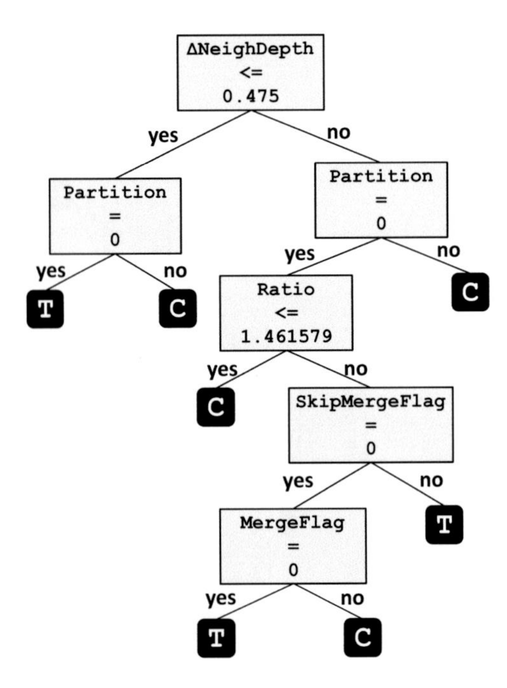

Fig. B.3 Coding tree early termination decision tree for 64×64 CUs

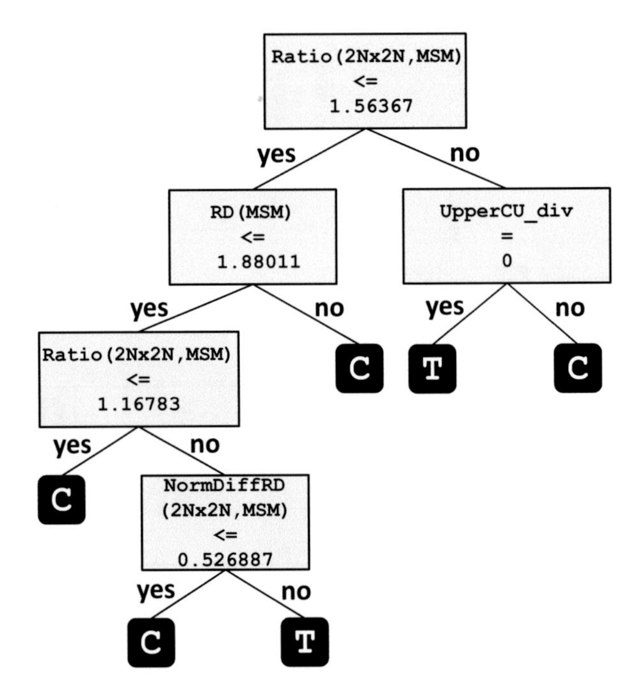

Fig. B.4 PU early termination decision tree for 8×8 CUs

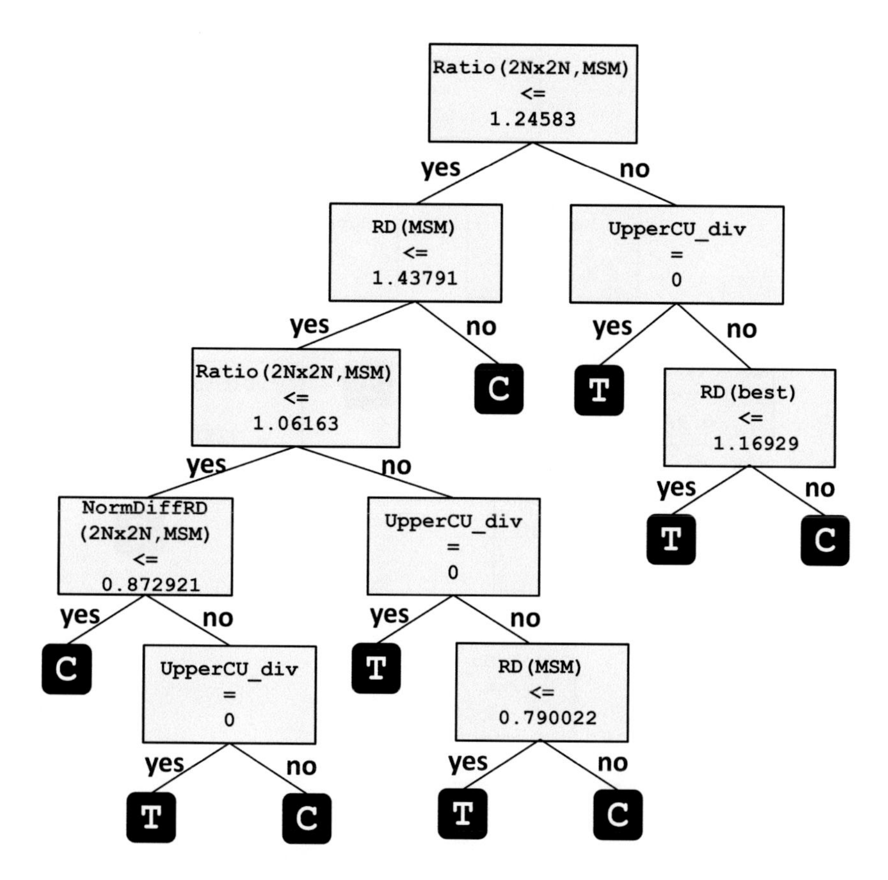

Fig. B.5 PU early termination decision tree for 16×16 CUs

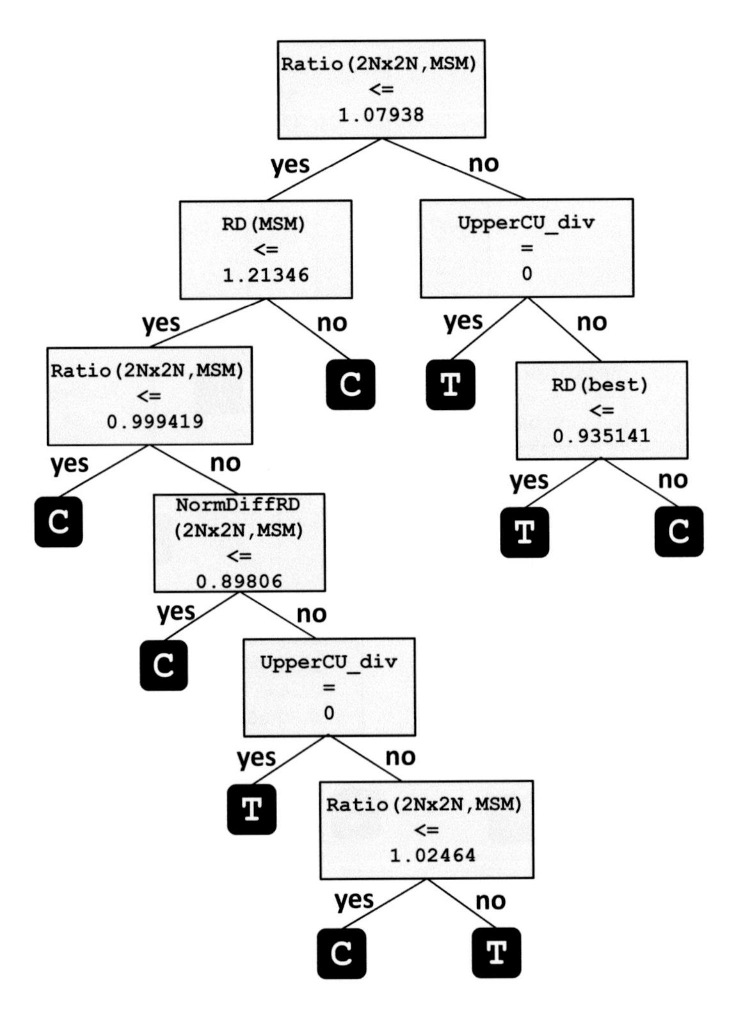

Fig. B.6 PU early termination decision tree for 32×32 CUs

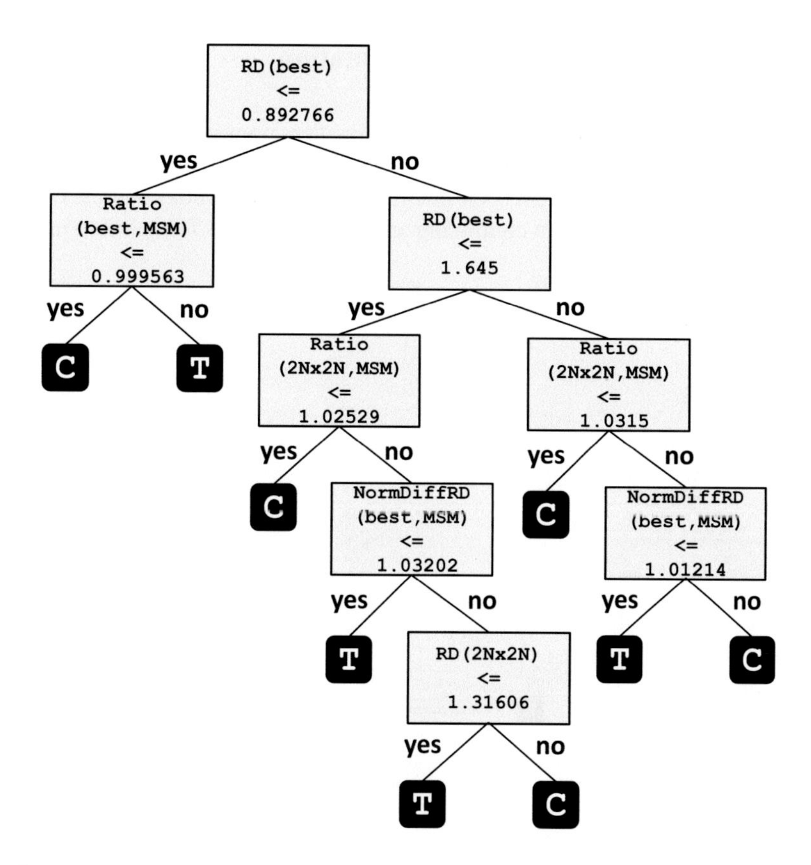

Fig. B.7 PU early termination decision tree for 64×64 CUs

Decision Trees for RQT Early Termination

Two decision trees for the RQT early termination were trained and implemented, as explained in Sect. 6.5. Figure B.8 presents the decision tree obtained for 16 × 16 TUs and Fig. B.9 shows the decision tree for 32 × 32 TUs. In both figures, *C* and *T* correspond to the decisions of continuing and terminating the TU splitting process, respectively.

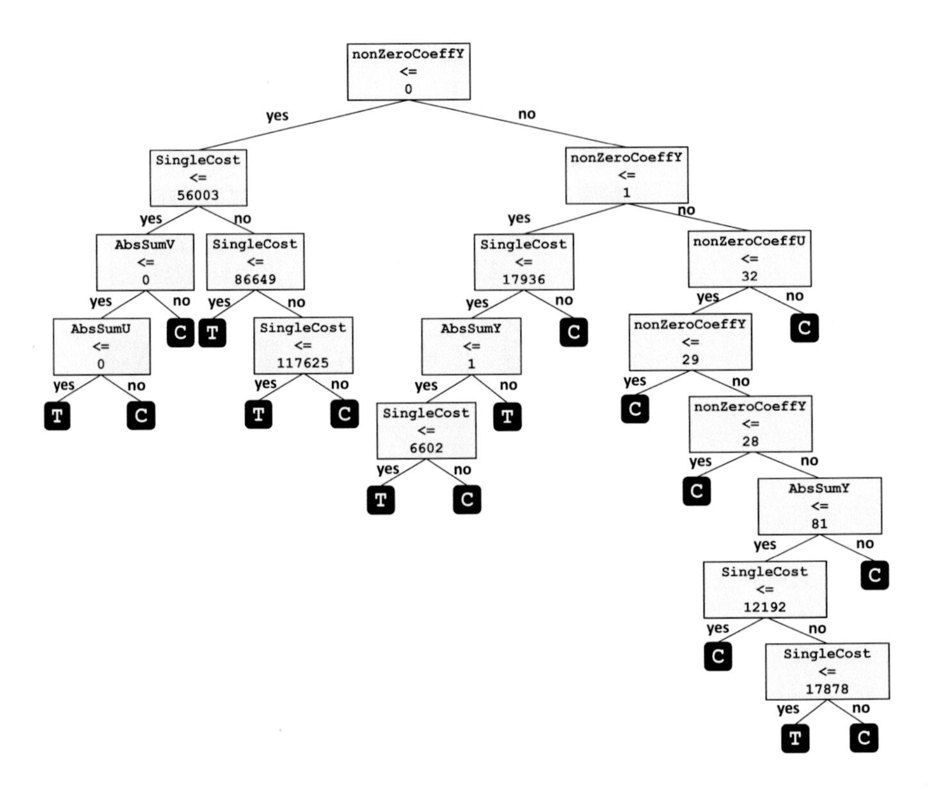

Fig. B.8 RQT early termination decision tree for 16 × 16 TUs

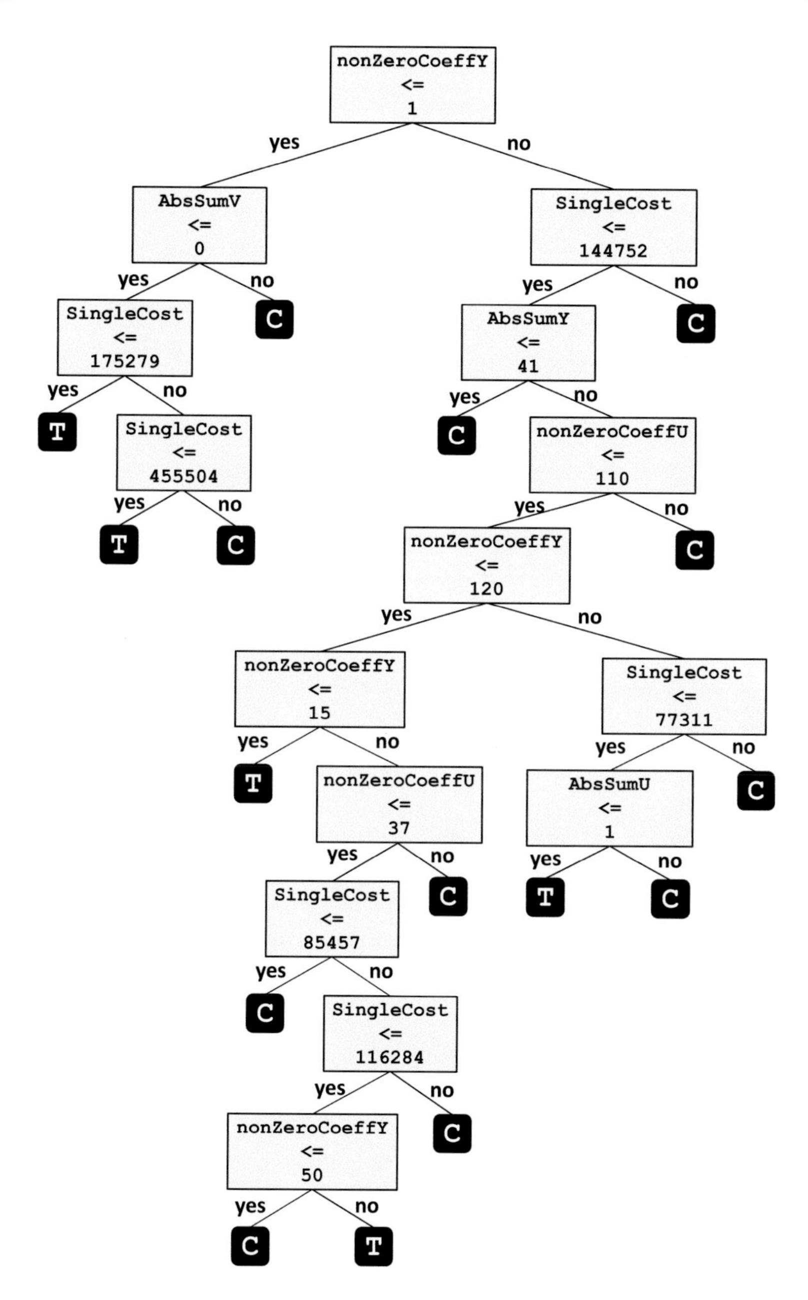

Fig. B.9 RQT early termination decision tree for 32×32 TUs

Reference

1. M. Hall, E. Frank, G. Holmes, B. Pfahringer, P. Reutemann, I.H. Witten, The WEKA data mining software: an update, *SIGKDD* Explor. Newsl, **11**, 10–18 (2009)

Appendix C
Encoder Configurations Tested in the R-D-C Analysis

In Sect. 7.1 it was explained that the encoding configurations considered in the R-D-C analysis were created by modifying the value of each parameter, one at a time, so that every parameter value could be tested with all possible values of the remaining ones, totalising 240 encoding configurations. As described in that section, the R-D efficiency and the computational complexity associated to each configuration was assessed with 10 high-resolution video sequences; QPs 22, 27, 32 and 37; and the *Random Access* temporal configuration, totalising 9,600 encodings. Average BD-rate, BD-PSNR and computational complexity reduction for each configuration, using the unmodified encoder as reference (configuration 1), were calculated. Each configuration tested and their respective results are presented in this appendix, in Table C.1, since only those corresponding to the points that compose the Pareto frontier were presented in Table 7.2 of Chap. 7.

Table C.2 shows the look-up table (LUT) used to determine the encoding configuration that best suits a given $R_{T(i)}$ ratio between the target time (T_T) and the weighted average encoding time ($T_{W(i)}$) of the last two GOPs, as explained in Sect. 7.2.1 of Chap. 7 (see (Eq. 7.1)). The encoding configuration used in the current GOP is used to select a line in the LUT where the closest value to $R_{T(i)}$ is searched. Once it is found, the index indicated by the column where the found value belongs is chosen as the new encoding configuration to be used in the next GOP.

G. Corrêa et al., *Complexity-Aware High Efficiency Video Coding*,
DOI 10.1007/978-3-319-25778-5

Table C.1 Parameters, computational complexity, BD-PSNR and BD-rate for the 240 encoder configurations considered in the R-D-C analysis

Config.	*SR*	*BPR*	*HME*	*CTET*	*PUET*	*RQTET*	Normal. Complex	BD-PSNR (dB)	BD-rate (%)
1	64	4	On	Off	Off	Off	1.000	0.000	0.000
2	32	4	On	Off	Off	Off	0.969	0.000	0.016
3	16	4	On	Off	Off	Off	0.958	–0.014	0.426
4	8	4	On	Off	Off	Off	0.950	–0.043	1.276
5	4	4	On	Off	Off	Off	0.945	–0.082	2.469
6	64	2	On	Off	Off	Off	0.973	–0.001	0.038
7	32	2	On	Off	Off	Off	0.941	–0.004	0.150
8	16	2	On	Off	Off	Off	0.930	–0.015	0.426
9	8	2	On	Off	Off	Off	0.923	–0.049	1.498
10	4	2	On	Off	Off	Off	0.917	–0.096	2.845
11	64	1	On	Off	Off	Off	0.964	–0.002	0.089
12	32	1	On	Off	Off	Off	0.933	–0.006	0.221
13	16	1	On	Off	Off	Off	0.922	–0.019	0.588
14	8	1	On	Off	Off	Off	0.915	–0.048	1.481
15	4	1	On	Off	Off	Off	0.910	–0.104	3.093
16	64	4	Off	Off	Off	Off	0.899	–0.018	0.595
17	32	4	Off	Off	Off	Off	0.868	–0.019	0.637
18	16	4	Off	Off	Off	Off	0.857	–0.029	0.886
19	8	4	Off	Off	Off	Off	0.849	–0.056	1.727
20	4	4	Off	Off	Off	Off	0.845	–0.100	3.045
21	64	2	Off	Off	Off	Off	0.872	–0.019	0.648
22	32	2	Off	Off	Off	Off	0.841	–0.019	0.651
23	16	2	Off	Off	Off	Off	0.829	–0.031	0.990
24	8	2	Off	Off	Off	Off	0.822	–0.059	1.839
25	4	2	Off	Off	Off	Off	0.817	–0.112	3.386
26	64	1	Off	Off	Off	Off	0.863	–0.020	0.697
27	32	1	Off	Off	Off	Off	0.832	–0.021	0.735
28	16	1	Off	Off	Off	Off	0.821	–0.033	1.067
29	8	1	Off	Off	Off	Off	0.814	–0.062	1.956
30	4	1	Off	Off	Off	Off	0.809	–0.120	3.690
31	64	4	On	On	Off	Off	0.723	–0.002	0.086
32	32	4	On	On	Off	Off	0.698	–0.003	0.124
33	16	4	On	On	Off	Off	0.688	–0.017	0.527
34	8	4	On	On	Off	Off	0.683	–0.045	1.337
35	4	4	On	On	Off	Off	0.680	–0.085	2.495
36	64	2	On	On	Off	Off	0.706	–0.004	0.145
37	32	2	On	On	Off	Off	0.680	–0.007	0.239
38	16	2	On	On	Off	Off	0.669	–0.017	0.550
39	8	2	On	On	Off	Off	0.664	–0.047	1.457

(continued)

Table C.1 (continued)

Config.	*SR*	*BPR*	*HME*	*CTET*	*PUET*	*RQTET*	Normal. Complex	BD-PSNR (dB)	BD-rate (%)
40	4	2	On	On	Off	Off	0.662	–0.102	3.051
41	64	1	On	On	Off	Off	0.700	–0.007	0.255
42	32	1	On	On	Off	Off	0.673	–0.006	0.256
43	16	1	On	On	Off	Off	0.666	–0.021	0.692
44	8	1	On	On	Off	Off	0.659	–0.052	1.633
45	4	1	On	On	Off	Off	0.657	–0.108	3.211
46	64	4	Off	On	Off	Off	0.660	–0.020	0.678
47	32	4	Off	On	Off	Off	0.634	–0.019	0.674
48	16	4	Off	On	Off	Off	0.625	–0.032	1.018
49	8	4	Off	On	Off	Off	0.618	–0.059	1.846
50	4	4	Off	On	Off	Off	0.616	–0.104	3.175
51	64	2	Off	On	Off	Off	0.643	–0.021	0.730
52	32	2	Off	On	Off	Off	0.616	–0.020	0.715
53	16	2	Off	On	Off	Off	0.607	–0.034	1.113
54	8	2	Off	On	Off	Off	0.601	–0.063	1.979
55	4	2	Off	On	Off	Off	0.598	–0.112	3.392
56	64	1	Off	On	Off	Off	0.637	–0.022	0.761
57	32	1	Off	On	Off	Off	0.610	–0.022	0.796
58	16	1	Off	On	Off	Off	0.601	–0.035	1.159
59	8	1	Off	On	Off	Off	0.596	–0.065	2.062
60	4	1	Off	On	Off	Off	0.594	–0.122	3.833
61	64	4	On	Off	On	Off	0.586	–0.018	0.572
62	32	4	On	Off	On	Off	0.569	–0.019	0.603
63	16	4	On	Off	On	Off	0.563	–0.029	0.847
64	8	4	On	Off	On	Off	0.559	–0.059	1.782
65	4	4	On	Off	On	Off	0.558	–0.099	2.946
66	64	2	On	Off	On	Off	0.572	–0.019	0.625
67	32	2	On	Off	On	Off	0.554	–0.019	0.620
68	16	2	On	Off	On	Off	0.548	–0.030	0.923
69	8	2	On	Off	On	Off	0.545	–0.061	1.916
70	4	2	On	Off	On	Off	0.545	–0.115	3.470
71	64	1	On	Off	On	Off	0.567	–0.019	0.611
72	32	1	On	Off	On	Off	0.551	–0.018	0.600
73	16	1	On	Off	On	Off	0.544	–0.034	1.044
74	8	1	On	Off	On	Off	0.541	–0.065	1.985
75	4	1	On	Off	On	Off	0.540	–0.118	3.534
76	64	4	Off	Off	On	Off	0.538	–0.032	1.001
77	32	4	Off	Off	On	Off	0.521	–0.033	1.074
78	16	4	Off	Off	On	Off	0.514	–0.044	1.392
79	8	4	Off	Off	On	Off	0.511	–0.071	2.205

(continued)

Table C.1 (continued)

Config.	*SR*	*BPR*	*HME*	*CTET*	*PUET*	*RQTET*	Normal. Complex	BD-PSNR (dB)	BD-rate (%)
80	4	4	Off	Off	On	Off	0.511	–0.115	3.484
81	64	2	Off	Off	On	Off	0.525	–0.035	1.122
82	32	2	Off	Off	On	Off	0.507	–0.034	1.094
83	16	2	Off	Off	On	Off	0.501	–0.048	1.545
84	8	2	Off	Off	On	Off	0.497	–0.075	2.322
85	4	2	Off	Off	On	Off	0.496	–0.124	3.757
86	64	1	Off	Off	On	Off	0.520	–0.035	1.171
87	32	1	Off	Off	On	Off	0.503	–0.037	1.242
88	16	1	Off	Off	On	Off	0.497	–0.047	1.507
89	8	1	Off	Off	On	Off	0.493	–0.077	2.446
90	4	1	Off	Off	On	Off	0.493	–0.134	4.179
91	64	4	On	On	On	Off	0.481	–0.026	0.823
92	32	4	On	On	On	Off	0.465	–0.027	0.865
93	16	4	On	On	On	Off	0.460	–0.039	1.186
94	8	4	On	On	On	Off	0.456	–0.066	2.013
95	4	4	On	On	On	Off	0.455	–0.108	3.305
96	64	2	On	On	On	Off	0.471	–0.026	0.832
97	32	2	On	On	On	Off	0.455	–0.027	0.879
98	16	2	On	On	On	Off	0.449	–0.042	1.291
99	8	2	On	On	On	Off	0.447	–0.071	2.219
100	4	2	On	On	On	Off	0.446	–0.122	3.708
101	64	1	On	On	On	Off	0.468	–0.029	0.958
102	32	1	On	On	On	Off	0.451	–0.029	0.939
103	16	1	On	On	On	Off	0.447	–0.041	1.294
104	8	1	On	On	On	Off	0.443	–0.073	2.305
105	4	1	On	On	On	Off	0.443	–0.128	3.910
106	64	4	Off	On	On	Off	0.445	–0.041	1.330
107	32	4	Off	On	On	Off	0.428	–0.041	1.339
108	16	4	Off	On	On	Off	0.422	–0.054	1.698
109	8	4	Off	On	On	Off	0.419	–0.081	2.553
110	4	4	Off	On	On	Off	0.419	–0.123	3.746
111	64	2	Off	On	On	Off	0.434	–0.043	1.408
112	32	2	Off	On	On	Off	0.419	–0.044	1.422
113	16	2	Off	On	On	Off	0.412	–0.056	1.821
114	8	2	Off	On	On	Off	0.409	–0.087	2.700
115	4	2	Off	On	On	Off	0.409	–0.136	4.217
116	64	1	Off	On	On	Off	0.432	–0.045	1.460
117	32	1	Off	On	On	Off	0.415	–0.044	1.443
118	16	1	Off	On	On	Off	0.409	–0.057	1.828
119	8	1	Off	On	On	Off	0.405	–0.088	2.795

(continued)

Table C.1 (continued)

Config.	*SR*	*BPR*	*HME*	*CTET*	*PUET*	*RQTET*	Normal. Complex	BD-PSNR (dB)	BD-rate (%)
120	4	1	Off	On	On	Off	0.405	–0.142	4.460
121	64	4	On	Off	Off	On	0.925	–0.006	0.187
122	32	4	On	Off	Off	On	0.894	–0.008	0.275
123	16	4	On	Off	Off	On	0.884	–0.018	0.527
124	8	4	On	Off	Off	On	0.877	–0.045	1.307
125	4	4	On	Off	Off	On	0.872	–0.087	2.592
126	64	2	On	Off	Off	On	0.897	–0.008	0.252
127	32	2	On	Off	Off	On	0.867	–0.009	0.312
128	16	2	On	Off	Off	On	0.856	–0.021	0.649
129	8	2	On	Off	Off	On	0.849	–0.050	1.514
130	4	2	On	Off	Off	On	0.844	–0.101	3.052
131	64	1	On	Off	Off	On	0.890	–0.008	0.291
132	32	1	On	Off	Off	On	0.859	–0.009	0.350
133	16	1	On	Off	Off	On	0.848	–0.022	0.690
134	8	1	On	Off	Off	On	0.841	–0.054	1.650
135	4	1	On	Off	Off	On	0.837	–0.109	3.331
136	64	4	Off	Off	Off	On	0.824	–0.022	0.722
137	32	4	Off	Off	Off	On	0.793	–0.023	0.770
138	16	4	Off	Off	Off	On	0.782	–0.037	1.198
139	8	4	Off	Off	Off	On	0.775	–0.064	1.972
140	4	4	Off	Off	Off	On	0.771	–0.107	3.286
141	64	2	Off	Off	Off	On	0.797	–0.023	0.762
142	32	2	Off	Off	Off	On	0.766	–0.024	0.796
143	16	2	Off	Off	Off	On	0.754	–0.037	1.250
144	8	2	Off	Off	Off	On	0.748	–0.067	2.131
145	4	2	Off	Off	Off	On	0.744	–0.118	3.669
146	64	1	Off	Off	Off	On	0.788	–0.025	0.835
147	32	1	Off	Off	Off	On	0.758	–0.026	0.898
148	16	1	Off	Off	Off	On	0.746	–0.040	1.340
149	8	1	Off	Off	Off	On	0.740	–0.069	2.215
150	4	1	Off	Off	Off	On	0.736	–0.123	3.787
151	64	4	On	On	Off	On	0.674	–0.009	0.283
152	32	4	On	On	Off	On	0.648	–0.011	0.369
153	16	4	On	On	Off	On	0.638	–0.022	0.716
154	8	4	On	On	Off	On	0.632	–0.051	1.539
155	4	4	On	On	Off	On	0.631	–0.093	2.747
156	64	2	On	On	Off	On	0.656	–0.010	0.338
157	32	2	On	On	Off	On	0.628	–0.012	0.424
158	16	2	On	On	Off	On	0.620	–0.024	0.745
159	8	2	On	On	Off	On	0.614	–0.058	1.806

(continued)

Table C.1 (continued)

Config.	*SR*	*BPR*	*HME*	*CTET*	*PUET*	*RQTET*	Normal. Complex	BD-PSNR (dB)	BD-rate (%)
160	4	2	On	On	Off	On	0.613	–0.105	3.148
161	64	1	On	On	Off	On	0.650	–0.012	0.415
162	32	1	On	On	Off	On	0.625	–0.014	0.482
163	16	1	On	On	Off	On	0.615	–0.026	0.871
164	8	1	On	On	Off	On	0.609	–0.055	1.747
165	4	1	On	On	Off	On	0.608	–0.115	3.493
166	64	4	Off	On	Off	On	0.610	–0.025	0.849
167	32	4	Off	On	Off	On	0.584	–0.027	0.924
168	16	4	Off	On	Off	On	0.574	–0.036	1.169
169	8	4	Off	On	Off	On	0.568	–0.066	2.062
170	4	4	Off	On	Off	On	0.566	–0.109	3.307
171	64	2	Off	On	Off	On	0.591	–0.026	0.900
172	32	2	Off	On	Off	On	0.565	–0.028	0.954
173	16	2	Off	On	Off	On	0.556	–0.040	1.329
174	8	2	Off	On	Off	On	0.550	–0.068	2.149
175	4	2	Off	On	Off	On	0.550	–0.121	3.662
176	64	1	Off	On	Off	On	0.586	–0.028	0.944
177	32	1	Off	On	Off	On	0.560	–0.027	0.939
178	16	1	Off	On	Off	On	0.551	–0.042	1.402
179	8	1	Off	On	Off	On	0.545	–0.071	2.292
180	4	1	Off	On	Off	On	0.544	–0.129	3.993
181	64	4	On	Off	On	On	0.539	–0.022	0.659
182	32	4	On	Off	On	On	0.522	–0.024	0.737
183	16	4	On	Off	On	On	0.516	–0.036	1.054
184	8	4	On	Off	On	On	0.512	–0.066	1.979
185	4	4	On	Off	On	On	0.512	–0.104	3.174
186	64	2	On	Off	On	On	0.525	–0.024	0.770
187	32	2	On	Off	On	On	0.508	–0.023	0.734
188	16	2	On	Off	On	On	0.502	–0.038	1.142
189	8	2	On	Off	On	On	0.499	–0.070	2.163
190	4	2	On	Off	On	On	0.498	–0.117	3.472
191	64	1	On	Off	On	On	0.521	–0.026	0.857
192	32	1	On	Off	On	On	0.504	–0.026	0.844
193	16	1	On	Off	On	On	0.497	–0.038	1.127
194	8	1	On	Off	On	On	0.494	–0.067	2.082
195	4	1	On	Off	On	On	0.494	–0.125	3.799
196	64	4	Off	Off	On	On	0.491	–0.038	1.256
197	32	4	Off	Off	On	On	0.474	–0.039	1.277
198	16	4	Off	Off	On	On	0.468	–0.052	1.673
199	8	4	Off	Off	On	On	0.465	–0.078	2.451

(continued)

Table C.1 (continued)

Config.	*SR*	*BPR*	*HME*	*CTET*	*PUET*	*RQTET*	Normal. Complex	BD-PSNR (dB)	BD-rate (%)
200	4	4	Off	Off	On	On	0.464	–0.123	3.742
201	64	2	Off	Off	On	On	0.477	–0.040	1.297
202	32	2	Off	Off	On	On	0.460	–0.039	1.278
203	16	2	Off	Off	On	On	0.454	–0.053	1.676
204	8	2	Off	Off	On	On	0.451	–0.082	2.582
205	4	2	Off	Off	On	On	0.450	–0.133	4.072
206	64	1	Off	Off	On	On	0.473	–0.043	1.387
207	32	1	Off	Off	On	On	0.456	–0.041	1.349
208	16	1	Off	Off	On	On	0.450	–0.053	1.711
209	8	1	Off	Off	On	On	0.447	–0.083	2.632
210	4	1	Off	Off	On	On	0.446	–0.141	4.372
211	64	4	On	On	On	On	0.447	–0.031	0.969
212	32	4	On	On	On	On	0.431	–0.032	1.018
213	16	4	On	On	On	On	0.425	–0.045	1.391
214	8	4	On	On	On	On	0.422	–0.072	2.214
215	4	4	On	On	On	On	0.422	–0.112	3.432
216	64	2	On	On	On	On	0.437	–0.034	1.092
217	32	2	On	On	On	On	0.420	–0.034	1.091
218	16	2	On	On	On	On	0.415	–0.045	1.425
219	8	2	On	On	On	On	0.412	–0.076	2.366
220	4	2	On	On	On	On	0.413	–0.125	3.762
221	64	1	On	On	On	On	0.434	–0.036	1.187
222	32	1	On	On	On	On	0.417	–0.034	1.133
223	16	1	On	On	On	On	0.412	–0.047	1.472
224	8	1	On	On	On	On	0.410	–0.078	2.439
225	4	1	On	On	On	On	0.409	–0.135	4.151
226	64	4	Off	On	On	On	0.411	–0.048	1.566
227	32	4	Off	On	On	On	0.394	–0.049	1.612
228	16	4	Off	On	On	On	0.389	–0.059	1.890
229	8	4	Off	On	On	On	0.385	–0.087	2.734
230	4	4	Off	On	On	On	0.385	–0.130	3.998
231	64	2	Off	On	On	On	0.400	–0.048	1.563
232	32	2	Off	On	On	On	0.384	–0.049	1.601
233	16	2	Off	On	On	On	0.379	–0.061	1.956
234	8	2	Off	On	On	On	0.375	–0.091	2.851
235	4	2	Off	On	On	On	0.374	–0.140	4.326
236	64	1	Off	On	On	On	0.397	–0.050	1.654
237	32	1	Off	On	On	On	0.380	–0.049	1.616
238	16	1	Off	On	On	On	0.375	–0.062	2.032
239	8	1	Off	On	On	On	0.372	–0.093	2.923
240	4	1	Off	On	On	On	0.372	–0.148	4.585

Table C.2 *Update table* with ratios between normalised complexities of encoding configurations

		Next configuration index														
		1	2	3	4	5	6	7	8	9	10	11	12	13	14	15
Current configuration index	**1**	**1.000**	0.945	0.890	0.852	0.841	0.821	0.649	0.620	0.582	0.516	0.460	0.443	0.435	0.433	0.425
	2	1.058	**1.000**	0.942	0.902	0.890	0.869	0.687	0.656	0.616	0.546	0.486	0.469	0.460	0.458	0.450
	3	1.123	1.061	**1.000**	0.957	0.944	0.922	0.729	0.696	0.654	0.579	0.516	0.498	0.488	0.486	0.477
	4	1.174	1.109	1.045	**1.000**	0.987	0.963	0.761	0.728	0.683	0.606	0.539	0.520	0.510	0.508	0.499
	5	1.189	1.124	1.059	1.013	**1.000**	0.976	0.771	0.737	0.692	0.613	0.547	0.527	0.517	0.515	0.505
	6	1.218	1.151	1.085	1.038	1.024	**1.000**	0.790	0.755	0.709	0.628	0.560	0.540	0.530	0.528	0.518
	7	1.542	1.457	1.373	1.313	1.296	1.265	**1.000**	0.956	0.897	0.795	0.709	0.683	0.670	0.668	0.655
	8	1.613	1.524	1.436	1.374	1.357	1.324	1.046	**1.000**	0.939	0.832	0.741	0.715	0.701	0.699	0.686
	9	1.718	1.623	1.530	1.464	1.445	1.410	1.114	1.065	**1.000**	0.886	0.790	0.761	0.747	0.744	0.730
	10	1.938	1.831	1.726	1.651	1.630	1.591	1.257	1.202	1.128	**1.000**	0.891	0.859	0.843	0.839	0.824
	11	2.176	2.056	1.937	1.854	1.830	1.786	1.411	1.349	1.267	1.123	**1.000**	0.964	0.946	0.942	0.925
	12	2.257	2.133	2.010	1.923	1.898	1.853	1.464	1.399	1.314	1.164	1.037	**1.000**	0.981	0.977	0.959
	13	2.300	2.173	2.048	1.960	1.934	1.888	1.492	1.426	1.339	1.187	1.057	1.019	**1.000**	0.996	0.977
	14	2.309	2.182	2.056	1.967	1.942	1.895	1.498	1.431	1.344	1.191	1.061	1.023	1.004	**1.000**	0.981
	15	2.353	2.223	2.095	2.005	1.979	1.932	1.526	1.459	1.370	1.214	1.081	1.043	1.023	1.019	**1.000**

Index

G. Corrêa et al., *Complexity-Aware High Efficiency Video Coding*,
DOI 10.1007/978-3-319-25778-5

P

Q

R

S

T

U

V

W

Y